Power and Penury

# Power and Penury

*Government, technology and science in Philip II's Spain*

DAVID C. GOODMAN
*Lecturer in History of Science and Technology*
*The Open University*

The right of the
University of Cambridge
to print and sell
all manner of books
was granted by
Henry VIII in 1534.
The University has printed
and published continuously
since 1584.

CAMBRIDGE UNIVERSITY PRESS
*Cambridge*
*New York New Rochelle*
*Melbourne Sydney*

Published by the Press Syndicate of the University of Cambridge
The Pitt Building, Trumpington Street, Cambridge CB2 1RP
32 East 57th Street, New York, NY 10022, USA
10 Stamford Road, Oakleigh, Melbourne 3166, Australia



First published 1988

Printed in Great Britain at the University Press, Cambridge

British Library cataloguing in publication data

Goodman, David C.
Power and penury: government, technology
and science in Philip II's Spain.
1. Science–Spain–History–
16th century 2. Technology–Spain
–History–16th century
I Title
509'. 46 Q127.S7
ISBN 0521-30532-2

Library of Congress cataloging in publication data

Goodman, David C.
Power and penury: government, technology, and science in Philip II's
Spain/David C. Goodman.
p.cm.
Bibliography: p.
Includes index.
ISBN 0 521 30532 2
1. Technology and state–Spain–History– 16th century. 2. Science
and state–Spain–History–16th century. 3. Philip II, King of Spain, 1527-
1598. I. Title.
T26. S7G66 1988
338.94606–dc 19 87-24630 CIP

ISBN 0 521 30532 2

MTL

# Contents

# Illustrations

# Preface

Philip II's Spain continues to fascinate. Since Braudel's justly renowned *The Mediterranean and the Mediterranean World in the Age of Philip II* with its stimulating portrayal of individual and society in Spain during the clash with the Ottoman Empire, other illuminating studies have appeared. In recent years there have been important studies of the monarchy's powerful military machine, its commerce with the Indies, and the state of the treasury; and fresh biographies of the king have put him in a new, more human light.

The theme of this book is the crown's involvement with technology and natural science. I have been drawn to this not only because it is largely untrodden ground, but also through the expectation that there would be rewarding source material amongst the documents of government in the Spanish archives. Apart from Philip's known interest in the sciences, there was his determination, and that of his advisory councils, to bring organisation into everything concerning the monarchy; it therefore seemed promising to investigate the crown's interest in technology in a period of war and expansion.

My research has only been possible through the assistance of institutions which have financed three periods of study in Spain. I am very grateful to the British Academy, the Royal Society and the Open University for grants awarded to me; and to the Open University for its generous provision of study leave. My thanks also go to the staff of several Spanish libraries and archives, where I found much kindness and cooperation. In particular I would like to thank Gregorio de Andrés, O.S.A., curator of the Instituto de Valencia de Don Juan, Madrid; and José Luis Rodríguez de Diego, archivist at the Archivo General de Simancas for allowing me access to the index cards he is currently accumulating for a continuation of the catalogue of the *Guerra Antigua* section of the Archive.

D.C.G.

June 1986

# Abbreviations

AGI:GIG – Archivo General de Indias, Seville, *sección de Gobierno: Indiferente General* with numbers of *legajo* and *documento*.
AGI:L – *ibid., sección de Gobierno: Lima* with number of *legajo*.
AGI:M y P – *ibid., sección de Mapas y Planos.*
AGI:P – Archivo General de Indias, Seville, *sección de Patronato* with numbers of *legajo* and *ramo*.
AGS:CJH – Archivo General de Simancas, *sección de Consejo y Juntas de Hacienda* with numbers of *legajo* and *número*.
AGS:DC – *ibid., sección de Diversos de Castilla*, with numbers of *legajo* and *número*.
AGS:E – *ibid., sección de Secretaría de Estado*, with numbers of *legajo* and folio.
AGS:GA – *ibid., sección de Guerra Antigua* with numbers of *legajo* and *documento*.
AGS:PR – *ibid., sección de Patronato Real* with numbers of *legajo* and *documento*.
ARP – Archivo del Real Palacio, Madrid.
BM Add.MS – British Library, department of manuscripts, Additional Manuscripts.
BNM MS – Biblioteca Nacional de Madrid, *sección de manuscritos.*
*Codoin* – *Colección de documentos inéditos para la historia de España* (Madrid, 1842–92).
IVDJ – Instituto de Valencia de Don Juan, Madrid; with *envío* and folio of manuscript.
MN:FN – Museo Naval Madrid: *Colección Fernández de Navarrete*, with numbers of *tomo, documento* and folio.
MN:SB – *ibid., Colección Sans y de Barutell (Simancas)*, with numbers of *artículo* and *número*.
MN:VP – *ibid., Colección Vargas de Ponce*, with numbers of *documento* and folio.

# Glossary

*Alcalde*: local magistrate.
*Alcaldes examinadores:* physicians appointed by the crown in fifteenth-century Castile to inspect apothecaries' shops and license medical practitioners; renamed '*protomédicos*' in the sixteenth century.
*Arroba*: weight equal to 25 pounds.
*Asiento*: a contract between the crown and financiers.
*Atarazana*: arsenal.
*Audiencias*: high courts of justice which in the Indies were also powerful politically.
*Azumbre*: a liquid measure slightly over 2 litres.
*Casa de la Contratación*: House of Trade, established in Seville in 1503 to administer the trade with the Indies; it also supervised cartography and navigation.
*Cédula*: a royal decree signed by the king.
*Consejo de Guerra*: the royal Council of War.
*Consejo de Hacienda*: the royal Council of Finance.
*Consulta*: a report sent to the king by one of his advisory councils.
*Contadores mayores*: accountants of the exchequer.
*Conversos*: Moors, or more commonly persons of Jewish descent, whose families had been converted to Christianity, usually by force.
*Corregidor*: the king's representative in the cities; known in Seville as the *asistente*.
*Cortes*: parliament in Castile or Aragón.
Ducat (*ducado*): unit of account equivalent to 375 *maravedís*.
*Escudo*: gold coin worth 340 *maravedís* or 10 *reales*.
*Estado*: Measure of length equal to about 6 feet.
*Fueros*: special laws or privileges of medieval origin recognised by the crown of Castile; the term can refer to whole regions like Aragón, Catalonia or Navarre, or to municipal guilds.
*Libra*: a weight very nearly equivalent to a pound of 16 ounces.
*Licenciado:* the title of a university graduate.

*Limpieza de sangre*: purity of blood, signifying the absence of Jewish or Moorish descent, formally required for university matriculation and entry to all public offices.
*Maravedí*: the basic unit of accounting, also circulating in various multiples as coins of copper–silver alloy. See 'ducat', '*escudo*', '*real*' for its relation to these units.
Mark (*marco*): a weight of 8 ounces.
*Médico de cámara*: the title of a physician appointed to attend the royal family.
*Moriscos*: Moors converted to Christianity by force and remaining largely unassimilated.
*Padrón*: official, standard, navigational chart of the Indies; prepared and kept in the *Casa de la Contratación*.
*Peso ensayado*: unit of currency of the Indies, rated at 450 *maravedís*.
*Procuradores*: representatives of the towns in the *Cortes* of Castile.
*Protomédico*: a chief physician of the crown appointed to license and supervise medical practitioners.
*Quintal*: a weight equal to 100 *libras*.
*Real*: silver coin worth 34 *maravedís*.
*Relaciones*: term designating the descriptive surveys of localities which were commissioned by the crown.
*Tonelada*: unit of measure of a ship's cargo capacity equivalent to two standard wine-casks of Seville ('*pipas*'); approximately nine tenths of the English 'ton'. In Vizcaya the *tonel* was employed, equivalent to one and a quarter *toneladas*.
*Veedor*: crown inspector; the term was applied equally to officials of the armed forces and inspectors at royal mining establishments.
*Zahorí*: a person, often a Morisco, professing to possess powers of detecting subterranean water or buried treasure.

Map 1 The Iberian Peninsula, showing localities discussed in the text.

# 1

## The occult sciences: the crown's support and controls

In July 1566 Martín de Ayala, archbishop of Valencia and former professor of philosophy at Alcalá, lay on his death-bed recording the events of his troubled life. He did not know exactly how old he was because his mother had forgotten the year of his birth; records of baptism were not regularly kept in Spain until after the Council of Trent. But he was certain about the day of his birth, St Martin's Day, in November, when the sun was in the sign of Sagittarius. And he believed that this constellation was the cause of the extreme difficulty experienced by his mother in giving birth, the cause of adverse effects in the countryside, and the cause of vexation throughout his life. This learned prelate was convinced that the predictions of astrologers were 'fully confirmed' by his own experience.[1]

In 1576 Diego Enríquez, serving as commander of the elite regiment of *tercios* in Philip's kingdom of Sicily, witnessed extraordinary scenes during an outbreak of plague in Trápani which altered his beliefs. The streets were full of people running towards churches, telling of a miraculous sweating of holy statues, which like the plague they saw as a sign of God's wrath. At first sceptical Enríquez was astounded to see the statue of St Sebastian sweating drops 'just like a body affected by great fatigue', and soon he was left in no doubt that two crucifixes, similarly oozing water, had miraculously cured two victims of the plague merely by being brought to them. Describing these events in a letter to the president of Sicily, the duke of Terranova, Enríquez said his reason for writing was that until then he had been 'very incredulous but from what I have seen I now readily believe'.[2]

Another example of contemporary beliefs: in Velilla de Ebro (Aragón) there was a church whose bell was supposed to ring without human agency; it was taken to be a miraculous warning from heaven of important events. The bell was alleged to have rung at the deaths of Philip II's father, mother and two of his wives. And when the bell rang again in June 1601 testimony of this miracle was given by local notaries, magistrates and by Acencio del Molino, a surgeon of Zaragoza[3], and interpreted as divine displeasure at the continued presence in Spain of Mohammedans.

These examples illustrate the acceptance of superstitious or occult beliefs at various levels of Spanish society. Spanish historians used to argue that Spain in the sixteenth and seventeenth centuries was commendably free from magic and the occult, that what little there was proceeded from rural ignorance rather than evil intention, and that superstition took no worse a form than the benign use of spells to cure rabies or remove locusts.[4] This rosy view has since been invalidated by historians such as Caro Baroja[5], who has demonstrated the widespread participation of Spaniards in occult practices, well-intentioned and otherwise. Early modern Spain was no different in this respect from France, Italy or Germany, where the predilection for superstition and the occult, in addition to orthodox religious beliefs, led the French historian Lucien Febvre to describe the sixteenth century as 'le siècle qui veut croire'.

'Superstition' is a word which the historian has to use with care; but it is not anachronistic here. In the sixteenth century its clear definition and usage were important not only for officials of state and church who sought to eradicate suspect beliefs and legislate against practices associated with them, but also for practitioners of occult sciences who were anxious to justify their activities as licit. In central Europe Trithemius and Agrippa had distinguished their 'natural magic', the excellent contemplation and application of nature's secrets, from reprehensible demonic magic and superstition. Agrippa regretted that the uninformed confused natural and black magic; and Paracelsus' alchemy was liable to bring from his enemies accusations of consorting with the devil. In Spain the same confusion tended to occur. Alejo Venegas, an author of philosophical and religious works, wrote:

> The art of magic is the art of sages. It is of two types: one is called 'occult philosophy' which is the science of natural secrets; the other is 'superstition' and involves an occult pact with the devil. There is hardly one good writer on the first who does not also dabble in the second. And so it is better to study neither one nor the other; more harm than good can come from Cornelius Agrippa's books on occult philosophy.[6]

## *1 Philip II and the occult*

Philip II, at the summit of Spanish society, had the power to influence opinion. How did he react to contemporary interest in the occult? This aspect of his intriguing personality has brought conflicting interpretations ever since his reign. The black portrayal of Philip in Protestant propaganda presented him as the incarnation of evil. When William of Orange, leader of the Dutch Revolt against Philip, had a price put on his head – he was eventually murdered – he accused Philip of 'diabolical machinations' and of conspiring to kill him by poisoning, which at the time was associated with sorcery. Completely different eulogistic portrayals appeared in seventeenth-century Spain, notably Baltasar Porreño's *Dichos y hechos del Rey D. Felipe II* (1628), which presented Philip as a man whose pure Catholic faith and prudent mind kept him free from all types of superstitious beliefs and offensive arts. Porreño tells how Philip dismissed an astrologer and tore up his treatise containing the horoscope and predicted life of the future Philip III, making it clear that such works were useless, vain and impious. And similarly how Philip ordered the publication of an astrologer's forecast for 1579 because its dire predictions had not come to pass, in order to expose the worthlessness of astrological divination and to show how little it deserved the attention of Christians. Porreño also relates how Philip frequently did the reverse of what predictions advised, such as purposely setting out on journeys on Tuesdays, supposed by some to be unlucky days.[7]

Porreño's hagiographic account, though based on unsupported anecdotes, has continued to be drawn upon in recent historical writing. When in 1914 an Augustinian monk published an astrological manuscript from the Escorial, a horoscope of Philip prepared around 1550 by Matthias Hacus, a physician and mathematician from northern Europe, the monk quoted from Porreño and commented that 'we can imagine Philip laughing at his horoscope and intentionally acting contrary to its predictions'.[8] Perhaps Porreño's biography also led Bratli, a Danish historian, in his study of 1911, to conclude that 'on one point in particular Philip proved himself intellectually superior to his age: he was not superstitious and was always contemptuous of the auguries and prophecies of astrologers'.[9]

Yet Philip did not tear up Hacus' prognostication, but kept it until his death, one of the hundred especially cherished books which he kept for himself after he had donated others to the Escorial library.[10] This, by far the most detailed of the several horoscopes of Philip known to exist,[11] predicted that his birth under the influence of Jupiter and Saturn would dispose him to be just in administration, brave, and studious; that these celestial influences, disposed by divine providence, would bring Philip's subjects great advantages through his strengthening of true religion and his opposition to usury. He would be loved not only by his own people but also by his enemies. There would be difficulties ahead and because Mars had been influential at the time of his birth Philip would have to be 'more of a warrior than his father, the Emperor'; but in the end he would achieve victory over false religions. He would marry three or four times, have two surviving sons, one excellent and the other a disgrace, and he would live to at least 60 and have a gentle end. There was a good deal of comfort here and perhaps that helps to explain why Philip kept it.

The inadequacy of previous accounts of Philip's attitude to astrology has been shown in Taylor's impressive essay[12] which argues for Philip's deep involvement in the occult. Taylor has demonstrated the correspondence of star diagrams in the astrological frescos of the Escorial library to Hacus' horoscope of Philip. He argues that the frescos were inspired by Juan de Herrera, Philip's confidant and certainly a devotee of occult philosophy[13], who succeeded Juan Bautista de Toledo as the principal architect of the Escorial. In fact Taylor interprets the Escorial as a magical design based on the circle, square and triangle, supporting this, less convincingly, with the reflection that Juan Bautista de Toledo, who designed the building, had lived in Naples, 'one of the chief centres of magic in Europe'. And he reveals that the foundation stones of the Escorial and later of its church were both laid at astrologically auspicious times, when Jupiter and Saturn, the two planets associated with Philip, were in conjunction. All of this, according to Taylor, had Philip's approval, and added to the large number of works on the occult in the Escorial library, establishes Philip's commitment to the occult.

Other parts of Taylor's argument are less persuasive, such as his view that 'the king's introspective character would fit in with an interest in astrology and the occult'. And although *Picatrix*, a magical treatise and one of many Hermetic works in the Escorial, recommended the wearing of black as an effective means of attracting the beneficial influence of Saturn, there is no reason to suppose that this was why Philip wore black – black dress was commonly worn by noblemen at the time. Taylor almost certainly exaggerates when he considers that Philip made Herrera his constant companion in order to be his Magus and perform occult services for him. Herrera's services were often divorced from magic, as in his design of giant rotating cranes to lift large stones from quarries and position them on the rising structure of the Escorial. There was no hint of magic in his explanation to Philip of their mechanism, merely a straight Archimedean account of the principle of the lever.[14] And although some occult philosophy entered the curriculum of the Academy of Mathematics in Madrid, which Herrera persuaded Philip to establish and which Herrera directed, much of the teaching was concerned with non-magical military engineering, navigation and cosmography.

Kubler, another historian of art, has entered the debate, rejecting Taylor's interpretation. He argues that the Escorial was not based on magical designs but on rational aesthetics, derived from Augustine's Platonic ideas of harmony and the correspondence of parts to the whole. And as for the astrological frescos, he denies that they were due to Herrera and attributes them instead to the Hieronimite monk José de Sigüenza, librarian of the Escorial and chronicler of its construction. This conclusion changes the interpretation because Kubler accepts Sigüenza's statement that the reasons for the astrological frescos were that a royal library 'must include every taste', and to show, in accordance with God's intentions, that men had nothing to fear from stellar influences because the power of prayer and repentance was greater. Kubler supposes that this also represents Philip's opinion and, using Porreño's anecdotes for support, arrived at the unacceptably simple conclusion that Philip was consistently hostile to astrology.[15]

Other sources, including some new evidence, help to clarify Philip's attitude to astrology; but ambiguities remain. Cabrera de Córdoba, the meticulous chronicler of the reign, told how Philip's court attracted diviners and tricksters.[16] Several astrological discourses were certainly sent by authors seeking profitable employment in the royal service, assuming that Philip would take them seriously. In 1577 someone who had served Philip in an unspecified way in Flanders sent his account of the significance of the recent comet which had startled Europe by its size and caused fear from the widespread belief that comets were harbingers of disaster. There were grounds for supposing Philip might be interested because for centuries comets had been associated with the death of princes, based on the doctrine of the pseudo-Ptolemaic *Centiloquium*. The astrologer assured Philip that the comet 'has no significance for His Majesty's health' but that its location portended the violent death of William of Orange. There was such 'antipathy' between the horoscopes of the two men that it was inevitable that there was war between them, and now the comet showed that 'God was with your Majesty'. He thought it confirmed the meaning of the disposition of the constellations when the Escorial was begun, that 'Your Majesty should take courage in subduing his enemies like an all-destroying ray'. And if Philip wished he would explain how the constellations indicated the decline of Mohammedanism.[17]

A book on the significance of comets was dedicated to Philip by Bartolomé Barrientos of the university of Salamanca; it pointed out that comets were 'not always' signs of disaster to come.[18] And in January 1573 Antonio Gracián, royal secretary, thought Philip should see two discussions of a comet, one by López de Velasco, cosmographer of the Indies.[19]

There is no indication of Philip's reaction to any of these discussions. But when Gracián wrote another letter to him on comets Philip's reply left a unique record of his views on an astrological matter. The sceptical Gracián, informing Philip that the *corregidor* (civil governor appointed by the crown) of Carmona had sent his thoughts on the significance of the recent comet, commented that the urge to prophecy had become so widespread that 'even *corregidores* were now afflicted with the disease', and added that in antiquity the Romans would have given the comet their full attention. Philip replied sceptically that 'because the Romans were not Christians they believed whatever they liked'.[20]

Another of the king's servants, Giovanni-Battista Gesio frequently supplied astrological advice on foreign policy. Gesio had come to Madrid from his native Naples in the 1560s and served Philip as an expert in cosmography and mathematics. He invented an undecipherable code[21] for the most secret parts of Philip's correspondence and spent several years in the 1570s at the embassy in Lisbon, arguing for Philip's rights in the disputes with Portugal on territorial possessions overseas. In 1578 after a decade of Spain's increasing military involvement in the Netherlands, Gesio informed Philip that the signs in the heavens clearly indicated the need for firm action with the rebels, for 'certain diseases required violent remedies if they were not to recur'. Therefore the king should concentrate his forces in Flanders into a single stronghold and mete out exemplary punishments to rebels. Astrology showed that in this way 'Your Majesty will be able to restore religion by arms, impose new laws and increase the royal patrimony with the wealth and estates of rebels'.[22] This would hardly have impressed Philip as novel and inspired counsel, since this very policy had already been recently implemented to great effect. In July 1577 Don Juan of Austria, Philip's governor-general of the Netherlands, had seized the citadel of Namur, concentrated his forces there, and the following January routed the rebel army at Gembloux. Two months later Gesio communicated his celestial wisdom to the king.

But it was the Portuguese succession which occasioned most of Gesio's astrological advice. Sebastian of Portugal, ignoring the views of Philip, his uncle, had proceeded with his disastrous crusade into Africa and was killed at Alcázarquivir (1578). The unexpected death of the young monarch brought the ageing Cardinal Henry to the throne and the prospect of a disputed succession. There were several contenders: Catherine of Braganza; Ranuccio Farnese; Antonio, prior of Crato, the illegitimate nephew of Cardinal Henry; and Philip II whose strong claim was through his mother, Isabella of the royal house of Portugal. In the midst of the impending crisis, Gesio sent Philip a lengthy discourse demonstrating that 'the inheritance of the Kingdom of Portugal belongs to Your Majesty by laws divine and natural' in which he argued from analogies based on the laws of geometry, perspective, music, hydrostatics, and above all on astrology.[23] Representing Manuel of Portugal and his descendants as planets, he drew a diagram to show how royal power was communicated like planetary virtues. He compared the female claimants to retrograde planets which neither received nor communicated virtue, and eliminated them also because females were 'created to obey and not to rule'. His argument eventually concluded that the Portuguese crown necessarily passed to Philip. And he later tried to strengthen this from recently observed celestial phenomena[24], the 'writing in the heavens' which was the visible sign of divine intentions. Eclipses seen in the meridian of Lisbon, Madrid and again at the moment of Cardinal Henry's death confirmed where the rightful succession lay. But Gesio warned Philip that 'it was an error to suppose celestial benefits come without effort, for the stars do no more than dispose and facilitate this; to win the gifts they promise, it is also essential to act to frustrate the designs of opponents'. He therefore urged Philip to take the necessary military and political action to secure what was his by right.

Philip used diplomacy to press his case in Lisbon and after the death of King-Cardinal Henry in January 1580 he used his military power to annexe Portugal, crushing the resistance of Antonio and his supporters. But there is nothing to suggest that Philip took any guidance from Gesio's astrological reflections; if anything the evidence suggests the reverse because, to his astonishment and dismay, Gesio was recalled from Lisbon in 1579 at a time when he felt 'it would have been better for me to have been present at the dispute and not sent away'.[25] Back in Madrid he continued to write to the king reassuring him that astrology guaranteed that Portugal would be his within months.[26]

Philip's attitude is not clearly revealed in the few legislative initiatives which touched on astrology. When in 1571 the *Cortes* of Madrid complained of the inadequacy of medical treatment and blamed physicians' ignorance of planetary motions, it pressed Philip 'to order that henceforth no physician may graduate from any university without the degree of bachelor in astrology'. Philip's response was to ask members of his council to deliberate with the universities 'to see what should be done'.[27] Historians have seen this as an indication of Philip's lack of interest;[28] but this may well be a misinterpretation: although no royal establishment of chairs in astrology was announced, a genuine intention on Philip's part to pursue the suggestion cannot be ruled out. Philip's protection of astrology might be evident in legislation at another session of the *Cortes* of Madrid at the end of his reign. This was a reenactment of Juan II's fifteenth-century decree prohibiting divination from 'words, the casting of lots, spells, the inspection of water, crystal or mirrors'. Philip ordered the introduction of more effective measures to secure its enforcement, to eradicate such 'divinations, spells and other superstitious things'; it is noteworthy that there was no specific reference to astrological divination.[29]

The evidence therefore points both ways and no simple statement on Philip's astrological beliefs will do. His interest fell short of total commitment and his reservations never amounted to consistent hostility; his was a middle position of restrained curiosity.

Philip's attachment to the occult works of the medieval Majorcan philosopher Raymund Lull is clearer and well known, though his motives remain obscure. Lull's principal work, the 'Great Art', promised his disciples a rapid acquisition of knowledge, not only of nature's secrets but also of religious mysteries; it served the burning passion of his life – to convert unbelievers, especially Mohammedans to Christianity. The Art consisted of manipulating letters symbolising attributes of God (goodness, power, eternity) and the parts of the cosmos (the signs of the zodiac, animals and plants, the four sublunar elements).[30] The Art would reveal the secrets of medicine and elucidate the deepest mysteries of faith, the Incarnation and the Trinity.

Apart from Lull's authentic works, many spurious alchemical and Kabbalistic treatises were attributed to him, adding to his reputation as a magician. Juan Vileta, a Lullist canon of Barcelona who advised Philip on where Lull's treatises could be found, identified Lull with the most famous sage of occult philosophy, Hermes Trismegistus.[31] There were ardent followers of Lull in several parts of sixteenth-century Europe. Apart from the strongly-rooted tradition in Majorca, where his doctrines continued to be taught at the university of Palma, a chair of Lullist philosophy had been created in Paris; in Prague Philip's ambassador Guillén de San Clemente was said to have boasted of his descent from Lull.[32] Herrera's interest in Lull's doctrines was pronounced. His library, richly stocked with Lullist treatises; his own Lullist discourse on the properties of the cube; and his plan in an early will to establish a Lullist teaching institute in his birthplace near Santander all show the degree of his enthusiasm. And Herrera's constant presence awakened the king's interest.

By 1577 Philip was taking steps to secure a rapid collection of Lull's writings for the library of the Escorial. When Arias Montano, the distinguished biblical scholar, was instructed 'to procure for the royal library all of Raymond Lull's works, or as many as I can find', he replied that this would best be done through the viceroys of Catalonia and Aragón, and that he would inform them where many of the treatises could be found and who owned them. Ecclesiastical authorities in Catalonia and Majorca were similarly approached.[33] By these means the Escorial library soon housed numerous Lullist and pseudo-Lullist works; they were read and annotated by Philip.

A contemporary record of Philip's enthusiasm comes from Juan Seguí, canon and later rector of the university of Palma, Majorca where he directed Lullist studies. In 1580, while accompanying Philip on his way to Lisbon to receive the crown of Portugal, he and the king discussed Lull. The king asked Seguí to write a brief account of Lull's life, which he at once began and completed in Lisbon.[34] Seguí himself saw Lull's Art as the way to Christian truth and of particular contemporary value 'when for our sins our Church is so infested with heretics'. He looked to the monarchy to create colleges from where students, taught Lull's Art and 'the tongues of the infidel in Persia, Malacca, Angola, Congo, Brazil and Peru' would go out, willing to serve God and die for the faith, so imitating Lull's example.[35] Nothing as ambitious as this was ever undertaken by Philip, but he did act on Herrera's advice to introduce Lull's philosophy at the Academy of Mathematics in Madrid; and for this purpose translations of Lull's *Ars magna* and *Arbor scientiae* were prepared by Pedro de Guevara, who put them 'in our Castilian language in accordance with Your Majesty's wish to facilitate the teaching of all sciences in your Academy'.[36]

Lull's philosophy was not beyond suspicion; in the fourteenth century it had been condemned by the Aragonese Inquisition and since then by several popes.[37] When Philip considered the introduction of Lull's doctrines at the college for Hieronymites within the Escorial, Gesio advised strongly against it. The grandeur and harmony of the Escorial would bring Philip epithets like 'the magnificent' or 'resplendent'; but 'Your Majesty's reputation would not be assisted by the introduction there of the so-called art and science of Lull, which is neither a science nor an art, because science teaches us to know by demonstration, and art has its principles'. Lull's doctrines were as false as the teachings of Mohammed. And Lull's promise of acquiring all sciences in a short time would attract 'only the pusillanimous', because 'practitioners of his art will know nothing in thousands of years'.[38] Gesio's opinion was completely at odds with Philip's.

Philip exerted pressure on Rome in an attempt to remove Lull's notorious reputation there, even to the point of seeking his canonisation. Why was Lull's philosophy so attractive to Philip? Did he see it as the key to universal knowledge, as a way to elucidate religious mysteries, or as a means of converting the infidel? These aims were not separable for Lull though they may have been for Philip. We know from the prologue to Seguí's life of Lull that Philip's favourite work of Lull was the *Blanquerna*; but unfortunately that does not help us to isolate any particular part of the doctrine for the usual blend of themes occurs. In this myth Blanquerna, after achieving wisdom through contemplation in a forest, becomes pope and teaches Lull's Art, promising that this would secure not only the mastery of theology, medicine and other natural science, but also the conversion of the Infidel.

In the Escorial library pseudo-Lullian works formed part of the large collection of treatises on alchemy,[39] in the practice of which Philip became involved. Venice was an important source of many manuscripts and printed books on alchemy and other occult science purchased for the Escorial. In April 1572 the king instructed his ambassador in Venice, Diego Guzman de Silva, to investigate the availability of rare books and manuscripts, 'chiefly of some antiquity, many of which, it is believed, can be found in that city or in other parts of its territory' and 'to acquaint yourself with persons who know about these things and inform me of all in detail'[40]. Book-purchasing soon began; between May 1572 and July 1573 the ambassador bought 167 Greek and 234 Latin manuscripts at a cost of 637 *escudos* from booksellers and individual collectors such as Matteo Dandalo.[41] Twenty-six alchemical manuscripts, in Latin and Italian, were bought from a German, Daniel Ulstath; these included the *Codicillus* and other spurious works attributed to Lull; texts attributed to Geber; works by Arnaldo de Villanova, John of Rupecissa and Christoper of Paris; and anonymous treatises on transmutation.[42] Two brothers from Corfu, Nicephoro and Michel Eparcho, supplied 64 ancient Greek manuscripts including an anonymous work listed as *liber de inveniendo Thesaurus hoc est de Alchimia*.[43] Cases of books continued to arrive at Alicante from Venice in 1579.[44]

Philip first became involved in alchemical experiments during his stay in the Netherlands (1555–9). He had been called there by his father, the Emperor Charles V, who in formal acts of abdication passed on to him the lordship of the seventeen provinces, and soon afterwards the crowns of Castile and Aragón. Philip's stay was prolonged by the recurrence of war with France; a French force attacked Douai in 1557. Philip also inherited the huge debts of the

Castilian monarchy, and now the demands of war made the financial crisis even more severe. In 1557 Philip suspended the payment of debts and, according to the reports of Michele Suriano, Venetian ambassador in Brussels, sought an alchemical remedy for his shortage of money; the imports of American silver, though significant, had not yet attained the huge quantities that flooded in after 1560, especially in the 1590s. Suriano said that Philip had first engaged an Italian alchemist, Tiberio della Rocca, and later a German, whose experiments in Mechelen, using 'an ounce of his powder' converted six ounces of mercury to six of silver; the silver stood the test of hammering but not fire. Suriano added that Philip was pleased with the result and wanted to pay his troops with this silver but the provinces wouldn't accept the inferior currency.[45] More details were revealed by the next Venetian ambassador, who reported that he had interviewed the German, who was called Peter Sternberg, and that this alchemist working under the supervision of Ruy Gómez de Silva, Philip's friend and councillor of state, was paid 2000 ducats. He added that the king was present at some of the experiments.[46]

Suriano described these attempts to manufacture silver as 'not very honourable' and thought that was why Philip had tried to keep them secret. The reputation of some sixteenth-century monarchs suffered by dabbling in alchemy. This happened to Philip's nephew, the Emperor Rudolf II, whose withdrawal from affairs of state was blamed on the diversions of alchemy and his other passions for the occult. And it also seems to have happened to Ernest of Bavaria, whose election as prince-bishop of Liège was soon regretted by his ecclesiastical electors and the population because he was neglecting government for alchemy.[47] Alchemy was frequently seen as frivolous or potentially sinful through its association with fraud. Gesio advised Philip that 'evil works are always blamed on their authors; nobles acquire infamy from alchemistical or false silver or gold'.[48] But after his return to Spain Philip continued to give patronage to alchemists, though he never neglected affairs of state in order to perform experiments himself.

His confidence in the transmutation of base into precious metals was shaken by the outcome of experiments, secret and prolonged, by unnamed alchemists working in 1567 in a house in Madrid under the supervision of Philip's optimistic secretary, Pedro de Hoyo.[49] The experiments were attempts to produce gold from copper, lead and tin. When the alchemists were asked by de Hoyo if they could make 'seven or eight millions in a year' they replied that they could 'even make twenty'. Philip too hoped for success and, impressed by Hoyo's account of the dedication of the alchemists, wrote 'although I am incredulous of these things I am not so of this; yet it is not bad to be [incredulous] because if it should not succeed one will then not be so disappointed'. He also wrote: 'we commend ourself to God, that in time it may turn out well; soon we shall see'. De Hoyo spoke of the alchemists' total confidence and added: 'pray God the tests may turn out as I desire'. To this Philip replied 'you say nothing about whether they have made silver, and if they do so it would be no bad thing if they proved it'.[50] The king's scepticism increased and after the results of a month's trials he declared that 'these transmutations do not satisfy me'.

After this Philip's reaction to promises of alchemical transmutation of metals became cool. When in 1569 Marco Antonio Bufale came from Rome to offer Philip his alchemical secret he complained that he had been kept waiting for three months without hearing if the king was interested. Although the delay was partly caused by the illness of the official concerned, the *contador* Garnica, it may also reflect the king's lack of interest.[51] And when Juan de Zúñiga, ambassador in Rome, wrote to say that an alchemist of repute was on the way to see the king, Philip replied that 'it could be a great hoax just like the effects of this science'; but added: 'if he arrives here notice will be taken of what he says'.[52] A similarly mixed response, scepticism without outright rejection, was given to Juan Fernández, an alchemist of Madrid. The correspondence relating to this has not yet been interpreted correctly.[53] What actually happened was that Antonio Gracián, the royal secretary, informed Philip that Fernández, after years of enquiry and considerable expense, had 'found the secret' he had been searching for; and that though Gracián himself viewed alchemy 'with derision' he thought 'Your Majesty would wish to see this marvel for himself, and if not, he and his pretended secret could be sent to those who can scoff at him'. Philip replied 'I have always regarded this as a hoax' but asked Gracián not to send the man away; instead he wanted the alchemist to set to work and say 'when he has brought it to completion which I am sure he never will'.

Wealth through transmutation of metals was one of the goals of alchemy which Philip no longer took seriously. But alchemy was traditionally supposed to offer another prize: the conservation of health; and here Philip's interest seems to have been strong and sustained.[54] In 1564 he appointed Francis Holbeeck, a Fleming, 'distiller of waters and essences' to work at Aranjuez; from there he supplied the king's physicians with medicines prepared from myrtle, marjoram and other plants.[55] Others continued to perform the same function at Aranjuez at the end of Philip's reign.[56] The alchemical nature of this work is evident from Giovanni Vincenzio Forte, a Neapolitan, whom Philip appointed to perform distillations in Madrid, and who said he fulfilled 'Your Majesty's frequent orders for preparing quintessences, according to the practice of Raymond Lull for the health of the human body'.[57]

But it was within the Escorial that the most elaborate arrangements were made to supply the royal family with alchemical medicines. There work began in 1585 to construct a suite of communicating rooms which were fitted out as laboratories with chimneys, furnaces and water supply. Here were mortars, presses, water-baths and various forms of specially designed apparatus for distillation on a large scale, including a giant tower, 20 feet high, made of brass and fitted with 120 glass alembics, which was said to produce 180 pounds of distilled essences in 24 hours. Potable gold, essence of cinnamon and numerous other vegetable and mineral medicines were prepared and stored in two rooms – one was not enough – of the pharmacy which adjoined the infirmary. In all there seem to have been no fewer than eleven rooms on two floors devoted to what a visiting physician described as the preparation of 'materials of great virtue'.[58]

Those who worked in the laboratory included Diego de Santiago, who saw alchemical medicine as the true experimental way of Lull and Paracelsus which revealed 'the occult in nature', and he contrasted this with the sterile book-learning of conventional medicine.[59] Richard Stanyhurst,[60] a Catholic exile from Elizabethan England, was another of Philip's alchemists at the Escorial. He worked there in the 1590s and wrote several treatises[61] for the king in support of alchemical medicine which he was convinced could cure diseases (gout, stone, consumption, leprosy and syphilis) otherwise incurable; he prepared potable gold medicines for Philip. He argued that the preparation of medicines for curing human diseases was the more important part of alchemy. His emphasis here was on 'human' to distinguish this from the cure of 'metallic' diseases, which was how alchemists imagined the process of removing 'imperfections' of metals such as copper, bringing them to the perfection of gold. To justify Philip's patronage of what Stanyhurst called 'this secret science' he referred to two of the king's ancestors: Robert of Naples, whom Lull had taught alchemy, and Philip, Duke of Burgundy, whose chivalrous Order of the Golden Fleece Stanyhurst took to signify the acquisition of alchemical gold.[62]

But Stanyhurst also believed in the reality of the transmutation of copper into silver and of mercury into gold. And the real 'purpose of his 'Toque de Alchimia' (1593), the 'touchstone of alchemy', intentionally brief and 'free from prolixity because of the serious affairs which constantly require Your Majesty's attention', was to help the king to distinguish impostors from those true practitioners who knew how to make precious metals. It could well have been an attempt to restore Philip's lost confidence. So Stanyhurst advised the king to be wary of alchemists who asked for rare or expensive materials, because 'great wisdom' and not great expenditure was needed – the materials required were cheap and common. He recommended tests on their secret powders 'when they are not around'; and inspection of their vessels, charcoal and all other implements, ensuring that no one was allowed to tamper with them, so that the trick of concealing gold in the hollows of the apparatus might be discovered. In addition the behaviour of the alchemist should be considered because alchemy was 'a special gift of God' and Christian philosophers were more likely to succeed than those given to vice. And as for impostors who used 'superstition and magical words, seeking not the power of nature but the assistance of the devil' they could be exposed by the use of 'holy water, by making the sign of the cross and by other holy rites employed by the Catholic Church against such incantations and superstitions'.[63] Impostors should be punished; but the true practitioners accorded royal favour. A king should be 'neither unduly credulous nor wholly incredulous'; one extreme would result in loss of money and ridicule, the other in failure to recognise the true work of the sincere alchemist, 'just as the Indies would remain undiscovered if no one had believed Columbus'.[64]

According to Cabrera de Córdoba, his secretary and chronicler of the reign, Philip refused the services of Pachet, a Morisco herbalist sent to him by the viceroy of Valencia to cure his gout, even though he had the reputation of healing incurable diseases. This was because the Morisco had been tried by the Inquisition for using a demon to collect herbs, and Philip did not want to get well 'by such bad means'.[65] But if on this occasion he kept away from black magic he readily resorted to medical treatment bordering on the magical or superstitious. When one of his maidservants was slow to recover from a fever, Philip thought she would soon get well by wearing 'a little chain of gold my sister sent her' and 'some bracelets sent by my niece'.[66] And when for his own health Philip was offered an opal with supposedly powerful virtues, he clearly showed interest and asked for information on its powers to be extracted from the appropriate chapters in Pliny's *Natural History* and Albertus Magnus' work on the occult virtues of gems and their engraved images.[67]

Philip's zeal for holy relics is well known. José de Sigüenza described how a mass of relics entered the Escorial after Philip had successfully negotiated their transfer from German monasteries and churches. Four large wooden crates filled with relics reached Milan after 'miraculously' evading the Calvinists; for 'God did not want the bones of saints to fall into the mouths of such ravenous dogs'. From Genoa the holy cargo was carried by sea to Barcelona, a passage 'divinely assisted' by the total absence of adverse winds. And when in May 1598 Philip first saw the relics in Madrid he 'adored them with great reverence', regarding them as a 'treasure compared to which his kingdoms were nothing'. Before they were deposited in the chapel of the Escorial many were applied to the king's 'eyes, mouth, head and hands' to relieve the pains of gout.[68]

Holy relics had been thrown out by the Reformers in Wittenberg, Zurich and Neuchâtel. And even in Charles V's Spain there had been critics of relics; the Emperor's Erasmian secretary Alfonso de Valdés, dismissing the alleged gravity of the violation of relics during the sack of Rome by Imperial troops (1527), thought 'a living body is worth more than a hundred corpses' and that 'relics are traps for idolatry'.[69] But the cult of relics was confirmed at the closing session of the Council of Trent (3–4 December 1563) and those who rejected it were condemned; the Council declared that through the relics of saints God bestowed many benefits on mankind.

Because this was the official doctrine of the Catholic Church, Philip's enthusiasm for relics, and perhaps even the way he applied them physically, cannot count as superstition. But his behaviour during the illness of his son, Don Carlos, seems more questionable. In 1562, during his residence at Alcalá, Don Carlos fell downstairs, injured his head and lay unconscious. He was soon attended by a succession of royal physicians and surgeons, including the great Vesalius. Bleeding, purgatives, ointments and trepanning were all tried without success. And then the despairing Philip, who was at the bedside, asked for the Franciscan friars of the neighbouring monastery of Jesus María to come in procession with the corpse of Diego de Alcalá, a fifteenth-century friar with a reputation for miracles. The corpse was placed above the unconscious prince, and the next night the prince had a vision in which friar Diego appeared with a cross and told him that he would not die. Don Carlos soon recovered; Philip thanked God and asked the Pope to canonise Fray Diego, and this was done in 1568.[70]

The point here, admittedly a fine one, is that at the time Diego was not yet a saint. While Trent had reinforced the cult of saints' relics, it also sought to eradicate 'superstitions' and other abuses associated with the cult. The emphasis was on saints' relics; other relics were not to be recognised without prior authorisation from bishops and ultimately from Rome.[71] Perhaps Philip's action could be seen as infringing this; but it brought no criticism.

The least known of Philip's incursions into the occult concern his employment of *zahorís*, a name derived from the medieval Arabic term for the planet Venus which was supposed to be a guide to the knowledge of occult things.[72] These men and often young boys[73], who were believed to be particularly sensitive in this respect, claimed to possess the powers of detecting various things buried in the earth: treasure, precious metals, corpses and springs of water. And it was in order to find water to supply the Escorial, the Pardo palace and the palace at Madrid that Philip paid for the services of two *zahorís*, a man and his son aged about eight.

Some time during his period of office (1556–68) as secretary of the *Junta de Obras y Bosques* (created to administer all the royal residences and already functioning in the previous reign) Pedro de Hoyo told the king about a boy 'who sees water, whether deep or shallow, under the earth, but only on days that are sunny and bright'. Philip's reply was enthusiastic. He asked for the boy to go out 'on the first sunny day' and 'show where the water lies and the way it is to be brought here . . . I think this will be good'.[74]

The boy, accompanied by his father and Blas Hernández, Philip's fountain-maker, was soon at work in the grounds of the Pardo, the Escorial and the Alcázar of Madrid. The optimistic predictions of plentiful subterranean water led de Hoyo to suggest that this might be the way 'to remedy the sterility of the area', and Philip to say 'he is the best *zahorí* I have had so far', revealing that there had been others. Philip as usual wanted to be kept informed of all that was taking place and asked that 'records be kept of whatever the boy has seen so that investigations can be made'.

But the results of digging were disappointing and de Hoyo soon wrote to Philip that 'it is clear that *zahorís* are not always right'. The diggers had either encountered impenetrable ground or were digging to considerable depths without success. After digging 22 feet de Hoyo said there were just two more feet to reach the level indicated by the boy, 'though he has already been wrong by half as to the depth, which does not surprise me because he is so young'. Philip replied 'See that they continue to the end since they have so little further to go'. And when after other unsuccessful attempts Philip wanted to know how the boy explained these failures, de Hoyo wrote back 'I asked the boy why he was wrong and he replied that the water is definitely there but at a great depth'. Blas Hernández was concerned that the water was too deep to be useful and Philip sent instructions that digging should not be continued beyond the point at which the exploitation of water became impracticable. There is no evidence that Philip's *zahorís* ever led the diggers to discover useful supplies.

At about the time that Philip was encouraging this occult search for water, a fourteen year old boy was sentenced by the tribunal of the Inquisition of Zaragoza to fifty lashes, six years' instruction in a convent, and a year's exile for claiming to possess the powers of a *zahorí*.[75] And when at his trial in 1576 another victim of the Inquisition, Amador del Velasco, readily admitted that he had practised judicial astrology, he was anxious to disassociate himself from the very activity in which the king had become involved: 'As for water I have extracted no little, and I have acquired knowledge of its situation, but I am no *zahorí*.'[76] The seriousness with which some contemporary clergy regarded the practices of *zahorís* is evident in the writing of Tomás Sánchez, the distinguished Jesuit and moral theologian, who said that since it was impossible for anyone to see through a body as opaque as the earth, *zahorís* could 'only acquire their knowledge through the power of the devil', a conclusion which he thought was supported by the association of their powers with certain days; their activities were therefore 'a mortal sin'.[77]

## 2 *The Inquisition's censure and other clerical opinion*

The royal Council of the Inquisition was the governmental agency most involved with the control of the occult sciences in Spain. Ever since its foundation in Castile by Ferdinand and Isabella the *Consejo de la Suprema y General Inquisición* had been independent of Rome and wholly subject to the Spanish monarch who appointed its officials. Philip II inherited an Inquisition which, from its centre in Madrid, organised the activities of numerous tribunals: thirteen in peninsular Spain, including the recently-created tribunal of Galicia (1562); Mallorca; the Canaries; two in America: Mexico, Lima; Sardinia and Sicily. At first occupied with the punishment of heresy among Christianised Jews and then with the eradication of Protestantism, the Inquisition broadened its scope in the early years of Charles V's reign to investigate and punish those engaged in magic and witchcraft.[78] This new responsibility was evident in the first Spanish Index of prohibited books (1559), prepared by Fernando de Valdés, the Inquisitor General. Nearly all of the books it listed were banned because of their Protestant content or simply because of the known Protestant beliefs of their authors (for example Leonhart Fuchs' botanical and medical works, and Gesner's *Historia animalium*); but it also included a general prohibition of books on 'necromancy and other superstitions'. Some of the listed works were suspect on more than one ground, so that it is not clear that magic was the cause of their prohibition; for example Reuchlin's *De arte cabalistica*, both magical and Jewish.

The Inquisition was in no doubt where its duty lay in magical practices involving the invocation of demons; this was consort with the devil and to be severely punished. But it was much less clear about its intervention in other activities related to the occult. In 1568 an inquisitor in Barcelona who had fined a woman for attempting to cure the sick by charms and incantations was reprehended for exceeding his duty.[79] And in the early 1580s while one inquisitor, Juan de Arrese, sent by the Valladolid tribunal to investigate the teaching of astrology at the university of Salamanca, called for its prohibition,[80] another, Juan de Mendoza Porres, recommended toleration of astrology, arguing that 'there was nothing improper in knowing our dependence on the heavens', since this encouraged love of God.[81]

If the Inquisition hoped to find clarification from the Doctors of the Church some differences in doctrine were also apparent here. Augustine's *City of God* had condemned astrology as an evil art designed to substitute the falsehood of astral destiny for the truths of human free will and divine providence; the positions of stars had not caused the greatness of Rome, nor were they responsible for any similarities in the medical history of twins born under identical constellations.[82] Aquinas had also taught that astrological prediction of 'the future activities of men' or of 'future chance events' was 'superstitious and wrong' and exposed the practitioner to serving the devil, because 'foretelling the future was the very opposite of an act of religion: instead of looking to God for instruction, knowledge was sought by a tacit or explicit compact with the devil'. But he qualified his criticism stating that 'if, from observations of the stars, predictions are made of future events which have natural causes, for example rain and drought, then this is neither superstitious nor sinful'.[83]

It may well have been Aquinas who influenced the Church's policy on astrology, since the same qualified disapproval was given by the Council of Trent and again in the enlarged Spanish Indices issued by the Inquisitor General Gaspar de Quiroga in 1583–4. Quiroga prohibited books on all types of magic including palmistry and judicial astrology 'which predict future events dependent on man's free will'; but permitted astrological predictions on matters relating to 'agriculture, navigation and medicine', 'knowledge of an individual's inclinations and physical qualities from the time of birth' and 'general events of the world'. And this was soon followed by Sixtus V's stern Bull against divination (1586), which broke with the patronage given to astrology by popes in the early sixteenth century but still fell short of complete condemnation. Sixtus warned that those who engaged in divination to find treasure and other 'things occult' put their souls in danger since these black arts stemmed from the devil. Astrologers were guilty of attempting to usurp the power of God who alone had knowledge of future events; they were guilty also of betraying the Christian faith by 'attributing so much power, efficiency and virtue to the stars' that from the moment of a child's birth they predict his course of life, future honours and wealth, health, quarrels and imprisonment' all of which were dependent on his unpredictable free will and not necessitated by stellar influences. Deuteronomy taught that God had created the stars to serve man, not to compel him; and man enjoyed the protection of guardian angels, whose power was immensely greater than the stars. Sixtus permitted astrological predictions which assisted navigation, agriculture and medicine; but it is unclear if he was also prepared to accept the discussion of individuals' inclinations allowed by Quiroga. The Bull condemned predictions of events dependent on free will 'even though those who make such forecasts may say or protest that they do not do so with certainty'. This and all other illicit arts of divination were to be diligently investigated by prelates and inquisitors; severe penalties were prescribed for practitioners and those who read or possessed books of this nature. Sixtus ordered his ruling to be proclaimed once a year in every parish church and to be published in the vernacular 'so that all may know it.'

Sixtus had no grounds to expect obedience in Spain, where not only the Inquisition but the Church was controlled by the monarch. The Spanish monarch was the undisputed head of the Church in America and since 1523 had wrested the right to nominate all prelates in Spain. In 1572 Philip II strengthened the crown's control over the Church even more by prohibiting appeals to Rome from ecclesiastical disputes in Spain; he also claimed the right to control the publication of papal bulls. It is usually said that Sixtus V's Bull was not published in Spain until the middle of Philip III's reign; in fact it was published in both Latin and Castilian, soon after its promulgation, as an appendix to a book on prophecy printed in Segovia in 1588.[84] There is however little solid evidence that the Bull had much influence. It is conjectural that it had anything to do with the temporary suspension (1586) of the chair of astrology at Valencia, which in any case was soon restored in 1591; at Salamanca astrology continued to be taught in the late eighteenth century. And contrary to Sixtus' instructions, Geronymo Oller, an astrologer and priest of Manresa (Catalonia) predicted the outbreak of war from a comet in 1607, a prediction which the Carmelite Marco de Guadalajara later described as 'soon confirmed'.[84a]

Nor did the Spanish Inquisition devote much of its time to the prosecution of those who infringed Quiroga's instructions.

Although its activity reached a peak in 1585–94, arrests by the Spanish Inquisition for occult practices during Philip II's reign were never more than a small proportion of the total number of cases investigated. The most recent research has shown that illicit occult practices formed only 1 per cent of the cases tried by the tribunal of Toledo in 1560–1600; the figure for Zaragoza is 5.5 per cent of the 6 000 offenders tried in 1540–1700; and though in Valencia offences in this category amounted to 14 per cent for the period 1530–1700, most of these occurred in the seventeenth century. It is estimated that of the 29 000 cases known to have been investigated by the Inquisition in 1560–1614 no more than 4 per cent concerned the occult; instead the Inquisition in these years concentrated on the prosecution of blasphemy and sexual immorality, accounting for as much as 50 per cent of all cases.[85] Those convicted rarely went to the stake; the usual penalties were confiscation of goods (automatic on arrest), fines, spiritual penance (masses or seclusion in a convent), ostracism and exile; severe corporal punishment in the form of flogging was common and, worse, in contrast to contemporary ecclesiastical tribunals in England, the Spanish Inquisition occasionally used torture.

The occult practices most prosecuted were judicial astrology, magical rituals for discovering treasure and magical medicine. Witchcraft was also punished, though here the Inquisition's attitude, often one of scepticism, moderated punishments and contributed to Spain's comparative freedom from the witchcrazes which affected the rest of Europe, especially after its investigation at Logroño in 1610.[86] Alchemists were generally not troubled by the Inquisition. Unlike astrology, alchemy had received little criticism from the Church[87] and was not mentioned by either Quiroga or Sixtus V. This freedom from controversy is sufficient to explain the non-intervention of the Spanish Inquisition and there is no reason to attribute it to 'the prestige of alchemy at Philip II's court'.[88] A rare prosecution of an alchemist did occur in 1603 when Gerard Paris, born in Aachen, educated in Antwerp and living in Madrid for nearly thirty years, was arrested by the Inquisition. But his offence was heresy; he used alchemy to interpret the mysteries of the faith. He thought he could 'prove' the Immaculate Conception by the alchemical process of 'conception' whereby a pure material after impregnation with another substance was recovered in its original purity. And he also explained hell and purgatory in terms of the alchemical separation of 'the soul and body' of metals.[89]

There is also record of controls over occult practices by ecclesiastical authorities in the Spanish Netherlands. Philip II had reorganised the dioceses there and given the bishops inquisitorial powers. In the new diocese of Ypres Bishop Simonius investigated a suspect medical practice used by Spanish troops in the Army of Flanders. Their wounds were being treated by healers who applied cloth and recited the words: 'by Christ, with Christ and in Christ, God the Father is omnipotent for you. May the power of the Faith, the wisdom of the Son and the virtue of the Holy Ghost clear this wound of all evil'. The bishop decided that this was superstitious and prohibited its continuation, judging in accordance with Catholic tradition that the application of natural remedies was necessary for the treatment of disease.[90]

The clergy frequently expressed strong individual views on the occult. Antoine Perrenot, Cardinal Granvelle, Philip II's minister and primate of the Netherlands is said to have employed an astrologer,[91] yet his correspondence expresses utter scepticism. In a letter to the Spanish embassy in Venice he enclosed a horoscope of the king and wrote: 'I find diverse opinions on the hour of birth of His Majesty, but the work of the best of astrologers is sheer vanity; the principles of his art are more imaginary than certain'.[92] And discussing with another correspondent an astrologer's forecast of a short reign for the recently-elected Sixtus V, Granvelle commented 'His Holiness is not yet 60 and robust, and therefore there is nothing in what the Italian astrologer says; and my view is that the whole of judicial astrology is a baseless vanity'.[93]

Other members of the clergy wrote treatises containing sections on the occult or even entirely devoted to the subject. These invariably contained stern warnings against meddling with the devil, commonly believed to be the inspiration of illicit magic and superstitious practices. Pedro Ciruelo's much-read treatise on the condemnation of 'all superstitions and sorcery', first published around 1530 and continuing to reappear in new editions in Philip II's reign, was intended to ensure that 'good Christians protect themselves from these [practices] more than from serpents and vipers'.[94] Christians had to be attentive and flee from practices in which theologians perceived a pact with the devil. Those who succumbed were traitors to God, as guilty as if 'they had fled to the land of the Moors to renounce their faith'.[95] The devil's great ambition was to be worshipped by men and to achieve this he used his vast knowledge of the natural world, disclosing to his followers where hidden treasure or water-wells might be discovered, and imparting knowledge of future events from his complete understanding of celestial motions; for when the devil was cast out of heaven he lost grace but not his immense knowledge.[96]

Ciruelo, an academic theologian of Salamanca, had sought to make priests more vigilant and to secure stronger penalties. The same intentions were evident in the treatise on magic by Martin del Río, a Jesuit and a magistrate in Philip II's service, who had taught theology at Louvain and Salamanca. Reflecting that alchemists did not know enough to achieve transmutations into gold and were consequently tempted to seek the help of the devil, he concluded 'it appears to be best if the practice of alchemy were made illegal and a capital offence, unless Princes choose to permit it'.[97] Las Casas supposed that good angels sent by God protected the officials of the Inquisition from demonic influence, so allowing them to continue their investigation of superstitions without being themselves harmed; he said arrested sorcerers had told of the ineffectiveness of their attempts to bewitch Inquisitors.[98] But there is evidence that some of the clergy participated in illicit occult practices. If we accept what Amador del Velasco said at his trial by the tribunal of Toledo, his teaching of judicial astrology and chiromancy in Burgos was financed by the archbishop, and his predictions encouraged by an Augustine preacher. He alleged that the bishop of Astorga had lent him a work on astrological prognostication and he named two canons of Zamora who practised the same astrology for which he was being condemned. He stated that he had examined the palms of bishops and even that of Cardinal Espinosa, a former inquisitor-general and for a time the chief minister of Philip II. And he revealed that an Augustine friar told him how to find buried treasure by making the sign of the cross and uttering certain words in Latin.[76]

Independent evidence exists of the involvement of the clergy in treasure-hunting. A friar was brought before the tribunal in Valencia for conspiring with an Italian friar and a virgin girl to find treasure; he had made a circle in the soil, burned incense, and implored angels to subdue demons guarding the treasure.[99] While ecclesiastical authorities had called on priests to be vigilant in their flocks, some of the clergy had themselves succumbed to the temptations of the black arts. Inquisitors would not have been surprised; they were well aware of the widespread fascination for the occult. The Inquisition censured del Río for writing a treatise which, while it condemned magic, discussed it in such detail that 'anyone who wanted to practise sorcery had only to read this'. When del Río replied that authors were permitted to describe heretical doctrines so that they may be more effectively opposed, the Inquisition responded:

> It is very well known what the difference is between heresy and sorcery; because it is clear that men are extremely attracted to knowledge of the occult. And because such marvellous things are done by the art of magic, men are curious to know how these things are done; this does not apply to heresy which may therefore be discussed without such risks.[100]

## *3 Lay opinion and popular beliefs*

The attractions of the occult in sixteenth-century Spain are clear enough. The prospect of immediate wealth through alchemical gold or divination of treasure; the hope of curing diseases for which conventional medicine had nothing to offer; and the peace of mind amulets might bring during the frequent outbreaks of war and pestilence were amongst the motives which led a motley clientele, both educated and illiterate, to practitioners of the occult. Those who supplied these services, sincere practitioners or opportunists and tricksters, found ready custom. But the credulity of the lay population was mixed with scepticism so that the practitioners sometimes had to run the gauntlet of lay ridicule as well as the risk of arrest by the Inquisition. And they also had to avoid infringing the law: secular authorities were empowered to deal severely with diviners, and in the case of magical medicine, the *protomédicos*, the crown's supervisors, hounded all such unauthorised practitioners.

Some of the men who were close to the king held sceptical views about alchemy. Gracián, the royal secretary is one example; the diplomatic correspondence of the reign shows that there were others. When Juan de Vargas Mexia, Philip II's envoy to Savoy, learned that the duke was paying out 600 ducats a month to some

Frenchmen to carry out alchemical experiments, he became suspicious. He reported that the French must be using alchemy as a pretext for something else in this strategically important territory, because 'in my opinion the alchemists of this time are no good'. And he soon sent another report that the French had left Savoy after a week, 'more or less fleeing' because they owed thousands of ducats.[101] Gian Andrea Doria, the Genoese commander of Philip's Mediterranean armada and a member of the Council of State, scoffed at Marco Bragadino's attempts to make gold in Venice believing it 'will soon end in smoke like that of all others in this profession; and if the miracle he pretends to perform were true and His Majesty possessed the secret, the delays and losses of the [silver] fleets from the Indies would hardly be felt. And Bragadino, without needing to show it to anyone, would be rich enough to wage war on the Turk and the whole world'.[102] After Bragadino's efforts had come to nothing Doria wrote: 'I am astonished that in that city [Venice] where there are so many serious and discriminating men they should have taken so long to free themselves from deception'.[103] The duke of Terranova, royal governor of Milan, wrote to the Spanish embassy in Venice: 'our predictions about Bragadino's golden millions have been confirmed and I am not surprised that those who associated with him should be ashamed, so firmly had they believed him'.[104]

But it was astrology which most interested the laity because of the all-embracing scope of its predictions which touched on the weightiest affairs of state as well as everyday events in the lives of individuals. Astrology and astronomy were more sharply distinguished at the time than is usually supposed. In the prologue to his treatise on judicial astrology, which he taught as professor of mathematics at Seville, Diego Pérez de Mesa pointed out that the ancients used the same name – 'astrology' – for the study of celestial motions and the prediction of their sublunar effects; but that 'the moderns call one of these "Astronomy", signifying a rigorous, demonstrative, mathematical discipline concerning the infallible law of the stars, and the other "Judicial Astrology" which permits judgements that are not certain'.[105] And in another treatise Juan Pérez de Moya, a mathematics teacher and priest, told his readers 'it is not my intention to discuss the significance of comets but only their generation'; he referred those who wanted the former to 'Ptolemy and other astrologers'.[106]

The predictions of Spanish astrologers reveal two interesting features. First they show the slight influence of inquisitorial or papal regulations, in the form of preliminary face-saving remarks acknowledging the authority of the Church and the dominion of human free will; what then followed was not always in strict agreement with ecclesiastical requirements. Second, in several of the prognostications there is a detectable strain of encouragement for Spain's arduous roles as an imperial power and champion of Catholicism, which may have served simply to reassure or please prospective noble patrons; its use as propaganda cannot be ruled out but that remains conjectural.

There is evidence that astrologers were prepared to go so far as to question the sovereignty of human free will. An anonymous Castilian manuscript[107] of 1605 shows hardly a trace of the recent ecclesiastical curbs on astrology. According to this author God at the Creation had given control of the universe to 'spiritual Intelligences or Angels', assigning one to each planet (Orphiel to Saturn, Raphael to Mercury, Samuel to Mars). Each of these intelligences diffused its powers and took its turn to dominate the universe for a period of 354 years and 4 months, conserving the world and humanity. This was not a universe dictated by fate, since it had been designed by God and was always subject to His will; but there was no chance either because the 'weight, number and measure' of the universe had been imposed by the original divine plan. Was humanity also subject to this deterministic universe? God had given man the freedom to choose between good and evil; but although in principle his free will could not be forced by the stars, in fact the constellations affected 'human bodies and, more often than not, free will, which failing through ignorance allows effects to come to pass without preventing them as it might'.[108] Therefore the notable events of biblical, ancient and recent history were all explicable in terms of planetary influence. When Gabriel, Intelligence of the moon, succeeded for the twentieth time to universal government, the following events testified to the fluctuating fortunes associated with its ascendancy: Anne Boleyn's death and the loss of the Catholic Church in England; Charles V's capture of Tunis and wars with the German Protestants; the foundation of the Order of Jesus; the death of Calvin; Alva's control of Flanders; naval victory over the Turks at Lepanto; the death of Sebastian in Africa; the sack of Antwerp; the plague of Milan and the earthquakes in Ferrara. And 'from what has been said and the examples given any good astrologer, physician or historiographer may conjecture the events of the future and regard them as certain'.[109] The only real admission of uncertainty concerned the prediction of an individual's life-span, which required knowledge, usually lacking, of the precise moment of birth. In such circumstances the author thought men would do well to imitate the example of Philip II 'who against all predictions lived longer than was expected because of the care with which he led his life.'

Astrological elections for medical purposes were not only sanctioned by the Church; they were firmly established in official medicine. Some Spanish physicians such as Juan de Carmona raised objections, but most, like physicians in Italy and Germany, regarded astrology as indispensable for the understanding of the causes of disease and for their judicious treatment. Llorenç Coçar, a Valencian physician, believed that 'the light and motion of the sun, moon and planets change the air by warming, cooling, drying or moistening it, and imparting occult qualities. Consequently the air we respire is changed, and physicians who are experts in astrology will know more about the causes of health and disease'.[110] Luis Mercado, one of Philip II's physicians, thought that the terrible ravages of the plague in Spain in 1598 were due to celestial causes. And in another discussion of plague, Francisco de Silva y Oliduera, a physician in Granada, explained that 'although it may seem difficult to followers of Aristotle and Plato that celestial bodies of such beauty could cause our putrefaction, yet they do so, not by immediately causing contagion but by disposing the air'.[111]

Revealing evidence of the hold astrology could have on physicians comes from the report on the illness and recovery of Philip II's son, Don Carlos, prepared by one of the royal physicians who attended him, Santiago Olivares. This stated that 'contrary to popular belief the prince's cure was no miracle', but was due instead to 'ordinary and natural remedies'.[112] But Olivares added that the prince's accident had been predicted many years before in a warning that 'Carlos, prince of Spain, will run the risk of a fall from a height or, less likely, from a horse'. His comment that 'although I regard most of astrology as a hoax there is something in those parts concerning nativities and revolutions of the year'[113] (i.e. predictions for an individual for the coming year) showed attachment to two pillars of astrology which had received a full assault from the Church.

In November 1577 the Spanish ambassador in Venice received news from Madrid of an extraordinary phenomenon in the night sky:

> A comet has been seen for five nights towards the meridian and astrologers say they have neither seen nor read of anything as large. It appears at dusk and seems almost as big as the moon. And it leaves a trail over a great part of the heavens or region where it travels and lasts until eleven at night. During this time there has been much anxiety and prophesying. The astrologers are in demand and until they say what course things will take no decisions are taken.[114]

And in the same month a correspondent in Antwerp writing to Granvelle about military preparations around Namur, described the comet and said that:

> A friend who knows about these things told me that the comet is in the sign of Capricorn and that Abutuazar [*sic*] wrote that such comets often signify dissension between princes and the abasement of religion. I leave you to consider whether this is happening here.[115]

This comet, one of seven to appear in Philip II's reign, startled observers throughout Europe[116] because of its brilliance and persistence, causing Lutheran clergymen in Wittenberg and Rostock to urge their congregations to repent, and Roeslin, a physician of Alsace, to declare the imminent end of the world. Tycho Brahe, whose accurate observations on the comet showed for the first time that comets were celestial bodies and not meteorological phenomena of the sublunar region as was generally believed following Aristotle, also joined in the discussion of the comet's significance. He believed that comets were supernatural formations produced by God and signals of impending calamities. And he thought this comet had special significance for Spain and her Empire because it appeared in Sagittarius, supposedly the sign which had influence over Spain; he predicted the death of the king and authorities, civil war caused by religious divisions – 'a just punishment for their oppression of numerous true Christians' – and, because the comet travelled north-east, evil acts by the Spanish in the Netherlands.[117]

In Spain Micón produced a prognostication on the comet; now lost, copies are known to have belonged to Juan de Herrera and the king's private secretary Mateo Vázquez.[118] Another prognostication, addressed to the king, reassured him that although 'many astrologers had judged from its position in Sagittarius that the comet threatened Spain' this was without foundation. This conclusion was based on demonstrating the irrelevance of this sign of the zodiac in the history of Spain: in AD 74 there had been a great conjunction in Sagittarius, yet far from experiencing difficulties Spain 'never enjoyed such happiness and peace, which lasted over 340 years until the coming of the Vandals and Suevi'; and in AD 714 [*sic*] the Moorish conquest afflicted Spain with 'its greatest adversity, because she was enslaved and some say 700 000 men perished, yet neither then nor for 300 years after was there any conjunction in Sagittarius'.[119]

The author accepted that 'no notable calamity has ever come to any land without prior indication by some natural sign', such as a comet or the recent arrival of whales at Antwerp after swimming the Scheldt, a clear sign of foreign invasion there. And he believed that 'each kingdom or province has a certain fortune, fate or occult property impressed by the heavens' because 'out of 105 kings of Scotland only 8 have died a violent death while in England no king has been free from rebellion, sedition or war' and Spain 'has at all times been subject to invasion'.[120] But no astrologer could predict the fate of 'an empire so extensive with so many nations and tongues as that of the most powerful King Philip our lord, whose banners and standards have crossed more than one third of the world, from Sicily to Cuzco and to the province of Quito, in which distance there are nine hours of difference, that is, when it is nine o'clock at night here it is twelve noon there, and if we measure its extent north to south it amounts to one quarter of the earth'.[121] No sign of the zodiac could dominate so vast an empire, and the only predictable event from the location of this comet was the death of the king's enemy, the Prince of Orange who was born under an inauspicious sign entered by the comet.

An exhortation to exert Spanish imperial power appeared in an undated astrological prediction[122] which may refer to another of the reign's comets. The reddish tail pointing towards Barbary was taken to indicate Philip's conquest of the Moorish homeland, and therefore the author urged that the *tercios* should be kept in a state of readiness. Zeeland too would fall to Philip 'either by agreement or by force'. France would suffer great destruction and England defeat or exhaustion. And the author continued: 'I am not surprised that Spain takes on so much, because the constellations of the Spanish and of His Majesty are those of courageous conquerors'. He thought the comet also indicated pestilence in Spain and he implored God 'to deliver us from this evil and direct it against the infidel, pagans, schismatics and heretics'. These enemies would be defeated by a great effort from Spain: 'The Spanish will be forced to conquer because others do not undertake that enterprise'. And similar optimism for Spanish victories was deduced from the path of the comet of 1596. This, the author said, had moved 'from S to N inclining towards the E; and since Spain lies to the S of England, and England to the E of Spain it seems to me that the effects of this comet will be brought to England by the Spanish – perhaps the death or capture of the Queen of England'.[123]

In his chronicles of the reign Cabrera de Córdoba described the comet of 1577 and noted that its 'effects', lasting until 1590, included true indications of war in Flanders and the change of reign in Portugal.[124] He even interrupted his narrative with a special note on portents and prodigies, regarding these as divinely caused and distinguishing them from 'ostents' which were illusions of sense.[125]

There was also well-reasoned scepticism in Spain about the significance of the comet. One author argued that comets had appeared without being followed by any remarkable event, while famine and pestilence had occurred without being preceded by any celestial sign. He thought astrologers had no grounds for associating comets with particular regions of the earth which they faced, and no basis for predicting the duration of their effects; a comet could have been concealed by the sun for days before becoming visible, and therefore its place and time of origin, used for some predictions, were not precisely known. Nor did he believe there to be any foundation for the common belief that comets signalled the death of princes. If this were true 'no king or prince would die until his signalled hour'; yet princes had died without the appearance of comets and comets had come without the death of any prince. And in lands 'where monarchs are chosen by election and the will of men celestial virtues have less influence'. The only grain of truth in the belief might be that the air was 'affected and inflamed' by a comet and that this change was first felt by 'the more delicate and tender bodies of kings'. But the same atmospheric change could just as well afflict 'subjects depraved by vice', while the virtuous could sustain themselves by moderation and care of their health. The wise man would not live in fear of comets, but trust in God and follow reason, seeking to learn more about the little-known nature of comets instead of succumbing to man's ignorant tendency to 'befriend novelties and mysteries'. Professors of astrology forgot how often their predictions had gone wrong.[126]

Sceptical opinions of the educated have inevitably left much more trace in the written record than those of the illiterate. But there is no real separation of educated and popular views on the occult in sixteenth-century Spain. They shared the same superstitions and consulted practitioners of the same type. The mischievous intervention of the devil in everyday life was accepted by both the learned and the ignorant. The inquiry into the witch-craze at Logroño revealed that the population believed that the 'devil had a pharmacy' and supplied witches.[127] This was no different from the belief of the distinguished *converso* physician, Andrés Laguna, that the devil deluded witches by providing them with hallucinatory ointments which caused them to imagine their flights to the Sabbath.[128] Spanish peasants believed in the evil eye; *higas*, amulets against its attacks, have a long-established tradition in the peninsula, and in the sixteenth century took a Christian form with figures of St James serving as handles.[129]

The Church tried without much success to secure its monopoly in the villages, seeking to correct local deviations from Catholic orthodoxy. Quiroga's Index of 1583 ordered that the exorcism of storms and demons was to be performed only with officially approved ecclesiastical manuals. The Church also opposed the excessive attention given by peasants to their village shrines or local calendar. Ciruelo blamed prelates for allowing missals and breviaries which contained verses indicating unlucky dates or hours.[130] In the remote Asturias local superstition and religious ignorance, attributed to inadequacies of the clergy, led a canon in 1568 to call for Jesuit missionaries to convert a population which he described as no better than that of the Indies.[131]

## 4 Conversos, *Moriscos and the occult*

Priests may have deplored the superstitions and religious ignorance of their flocks in the outlying regions of Galicia and Asturias; but it was the newly Christianised Semitic minorities who, in the eyes of the Old Christian population, seemed most addicted to superstition and the black arts. These were the New Christians, the *conversos* (usually signifying converts of Jewish descent) and Moriscos (converted Moors), descendants of the Jews and Arabs who had made medieval Spain the land of magic, and whose conspicuous presence continued to distinguish Spanish society from the rest of Europe. Their reputation for magic was due in part to a misunderstanding of unfamiliar rituals, but they also had the legacy of a rich magical culture, parts of which continued to be influential in the reign of Philip II. Their activities were closely watched by the authorities.

Although the Old Testament prohibited sorcery and divination, later texts which were equally fundamental for Jewish religious life offered some support for magical beliefs and practices. The Talmud, with its multitude of rabbinic opinions, included demonology and statements of belief in the use of amulets for healing; and while maintaining the sovereignty of free will, some of the recorded rabbinic views sanctioned the belief that human character was influenced by the constellations at birth. The mystical Kabbalah, whose basic text, the Zohar, was written by rabbis in late thirteenth-century Spain, portrayed a world populated by spirits and demons; some of its devotees sought to control the powers of nature by occult script representing the signs of the zodiac. And in the *Shulchan Aruch* (Venice, 1564–5),[132] the final code of Jewish ethics, composed by a rabbi of Spanish origin and which has ever since continued to guide the orthodox, magic is allowed when it assists recovery from disease.

Apart from these religious works gentiles supposed that books on magic had been written by Moses, Daniel and especially Solomon, the king of wisdom. An alchemical text entitled *Idea Salomonis* was bought in Venice for the Escorial library.[42] And the spurious *Clavicula Salomonis*, which purported to be Solomon's legacy to his son Rehoboam, circulated in sixteenth-century Spain in spite of its prohibition by Quiroga's Index. The *Clavicula* gave directions for conjuring spirits and for subduing gnomes guarding buried treasure; it described various pentacles, magical diagrams inscribed with Hebrew letters and divine names, which were supposed to bring recovery from disease, victory in war, protection from sorcery and 'all other earthly dangers', and to instil knowledge of the virtues of all herbs and stones. A copy of the work was owned by Alonso de Verlanga, familiar of the Inquisition in Valencia; he was accused of using it to invoke demons.[133]

Those envious of the remarkable success of the Jews in Spanish finance and medicine were therefore able to attribute this to their knowledge of black magic. This seems to have happened in the early sixteenth century to López de Villalobos, *converso* and royal physician to Ferdinand the Catholic, when he was imprisoned by the Inquisition for eighty days on a charge of acquiring his exalted position by black magic; he was subsequently exonerated. And when in the seventeenth century Olivares the Count Duke, the dominant power in the early years of Philip IV's reign, fell from office, this minister of *converso* descent was accused of using magic to secure influence over the king and bringing a Jewish sorcerer from that haven of Spanish Jewry, Salonika.[134]

The circumstances of the conversions gave grounds for questioning the sincerity of the New Christians. *Conversos*, first appearing after the massacres of Jews which occurred in several Spanish cities in 1391, had been converted by force; and when in 1492 the Jews had been given the choice of baptism or expulsion, those who decided to remain in the peninsula included apparently sincere converts and others who continued to retain Jewish customs under a mask of Catholicism. The investigation of false *conversos* had been the original reason for the creation of the Spanish Inquisition (1478–80). The clergy were divided, some accepting that after baptism the Jew was no different from any other Christian. But racialist views were conspicuous in men like the friar of Toledo who wrote that while former Jews had become pious Christians they could not rid themselves of their 'evil inclination any more than the negro could his colour'; other clergy successfully secured the exclusion of *converso* wet-nurses from the royal palace because their milk might transmit evil characteristics.[135] Anti-semitic propaganda with accusations of malevolence was also spread by Juan Siliceo, once Philip II's tutor and since 1546 archbishop of Toledo. Calling on

Old Christians to undertake a second *reconquista*, this time against the enemy within – the *conversos* – he alleged that these had infiltrated the upper ranks of the clergy and that through their domination of Spanish medical practice they held the lives of Old Christians in their hands. He warned that a *converso* physician had been found to have a fingernail long enough to conceal poisons, which were administered by stirring with the finger purgatives given to his Christian patients; and he told of a *converso* surgeon who had been punished by the tribunal of Toledo for spreading powdered poison in the wounds of Old Christians. The archbishop had appealed to Charles V, without success, to introduce a regulation prohibiting *conversos* from practising medicine, surgery or pharmacy.[136]

Fears in Spain concerning the intentions of Jewish physicians had been expressed since the Middle Ages. In 1491 the grotesque myth of Jewish ritualistic killing of Christian children had been invoked at an Inquisitional trial for conspiracy with a Jewish physician, who was supposed to have asked for the heart of a boy for a magical attempt to inflict Inquisitors with rabies.[137] Accounts of the mythical poisoning of Henry III of Castile by his Jewish physician were still repeated in the sixteenth century. Yet *converso* physicians continued to be employed because of their skills and the shortage of Old Christian physicians.

Like the Jews, the Moors of Spain had been forcibly converted by mass baptisms beginning in Castile (1499), a few years after the conquest of Granada, and continuing in Aragón (1526). Their mosques had been converted into churches, their religious rituals (ablutions, circumcision, method of killing animals for food) prohibited, their literature and music banned; and it was only through payments to Charles V that they had managed to escape Ferdinand's earlier prohibition of their distinctive dress. These attempts at assimilation had been motivated by fears of renewed invasions from their African brethren, and the issue became increasingly urgent in Philip II's reign because of Morisco collaboration with their Turkish co-religionists. But they clung to their customs and were far less assimilated than the Jews. In the mid sixteenth century the Moriscos were concentrated in Granada, where they formed over half of the population, and in Aragón, about 20 per cent, chiefly in the kingdom of Valencia. In contrast to the converted Jews the Moriscos were not prominent in the professions; they were mostly artisans, pedlars and cultivators of irrigated land.[138]

The Old Christians regarded Moriscos and the Moors of Barbary with equal contempt, and viewed their attachment to the occult with a mixture of awe and condemnation. When the king's secretary was told of a 'very cunning' alchemist who might be of service, the individual referred to was a captured Moor.[139] The Moriscos according to the soldier and scholar of Arabic literature Hurtado de Mendoza (1503–75) were 'a nation as much given to astrological vanity and divination as their ancestors the Chaldeans'.[140] The Morisco witch is a recurring figure in Spanish literature. At the synod of Guadix (1554), convened above all to deal with Morisco vices, various of their occult practices were condemned: their casting of spells, divination and prophecy, use of *zahorís*, and burning of incense to cure attacks of the evil eye. Sorcerers were to be whipped and exposed to public shame at the doors of churches; those who consulted them were to do public penance in mass, standing with a rope around the neck.[141] And the later synod of Granada (1572), condemning Morisco superstitions, threatened with excommunication anyone who attempted to cure diseases with materials devoid of natural virtue.

These allegations were well-founded. Although Islam prohibited sorcery, magic was a conspicuous part of Morisco culture, as scholars conversant with their literature have been able to show. Their clandestine texts written in *Aljamía* (Castilian written in Arabic characters) include magical treatises with directions for conjuring genii; lists of the marvellous properties of the organs of certain animals; magical cures of the sick and the bewitched; magical uses of the verses of the Koran; and astrologically propitious times based on the consideration of the daily influence of stars and tutelary genii. Amulets are especially prominent; directions are given for their construction and use. One text contains the dictum: 'he who fails to protect himself with amulets is at the mercy of demons, just as a house without doors is open to all'.[142] The Koran itself contains special verses (Suras cxiii and cxiv, known as the 'preservative chapters') for protection against Satan, genii and the mischief of men and women; such verses were engraved on Morisco amulets.

Recent research on the medicine practised by Moriscos reveals their magical beliefs. In Aragón the Moriscos were attended by itinerant Morisco healers and herbalists, often farmers or shepherds with a rural familiarity with plants; quacks who conjured demons;

and fakirs, religious leaders who also treated disease by writing Arabic letters on parts of the body or administering paper, sometimes soaked and swallowed by the patient, inscribed with texts from the Koran. Maria la Pena was one of several female advisers on contraception; she prescribed swallowing three hairs in a little broth and the recitation of certain words.[143] At Cuenca Moriscos determined whether or not a disease would be fatal by studying the patterns assumed by a spoonful of molten lead, poured into cold water in a pot placed on the invalid's head.[144]

There were Old Christian fears that the Morisco physicians, like the *conversos*, were plotting to kill them. At Huerto (Aragón), Luis Comor was believed to have used his medicines to kill over twenty clergy.[143] These fears assisted the wider application from the mid sixteenth century of the statutes of *limpieza de sangre* which kept the Moriscos out of the universities and the medical profession. It has been suggested that it was this isolation from academic medicine, along with the suppression of Arabic culture with its medical tradition, which reduced Morisco medicine to the level of healing and quackery, an inferior medicine combining empirical herbal remedies with amulets, spells and the burning of incense.[145]

The numerous Moriscos arrested by the Inquisition were usually charged with observing a Mohammedan rite, but a few were specifically condemned for practising magical medicine with the assistance of demons. They were severely treated; some like Román Ramírez at Cuenca and Jerónimo Pachet at Valencia did not survive their torture and imprisonment. Controls were also introduced on Morisco midwives (Guadix 1554; Valencia 1561); their assistance at birth was permitted only in the presence of an Old Christian physician or surgeon who was instructed to prevent the performance of the 'superstitious ritual' of circumcision.[146]

Increasing ecclesiastical intolerance and renewed Ottoman belligerence in the Mediterranean persuaded Philip II to issue an uncompromising royal decree (1567) depriving the Moriscos of their traditional dress, their baths, their characteristic surnames, their Arabic speech and ordering the surrender of all Arabic books for examination. This led to the uprising of the Granadine Moriscos in the Alpujarra mountains (1568–70), a ferocious war which ended in their defeat and exile to other settlements in Castile; and in 1582 Philip's Council of State decided to rid Spain of the Moriscos once and for all, a decision ruthlessly executed (1609–14) in the next reign.

During the uprising prophecies incorporating legend and extraordinary phenomena circulated amongst the rebels. And several contemporary observers were convinced that these occult predictions had played an important part in inciting and sustaining the rebellion. According to Hurtado de Mendoza, Moriscos noted the monstrous births of animals in Baza, strange birds in Granada, apparitions of armed men in the skies above the Sierra Nevada, and judged from recent solar eclipses that this was an inauspicious time for Christians, whereas 'they believed themselves subject to the moon' and therefore unaffected.[140] Another historian of the war, Luis de Marmol Carvajal thought the rebel leaders had purposely played on Morisco superstition by inventing prophecies to persuade the people that the time had come to rise up and achieve their liberation, and 'so effective was this that even those who had invented them came to believe that things would come to pass just as they had predicted'.[147]

Some of the prophecies were discovered amongst the Arabic books seized by the Inquisition; another was found by a soldier in a cave in the Alpujarras. Their content was revealed through the translations of Alonso del Castillo,[148] the son of a Morisco, but an apparently sincere Christian who was employed as a translator by the Inquisition. One of the prophecies told that a comet 'very large and bright' would be the first sign of the Turkish conquest of the West and the return of Spain to Islam. Another, allegedly made by Ali, son-in-law of Mohammed, spoke of the liberation of Andalusia in a year 'which begins on a Saturday', a year 'full of mists and little rain, when the trees will bear much fruit, the harvests will be more abundant in the cool hills than on the coast, and the bees fill their hives'. A third prophecy was a letter from a hermit of Algiers to a rebel leader conveying his dream of Morisco victory.[149]

But what is most interesting, and generally overlooked, is that prophecy was used as a strategem of war by the royal authorities to weaken the morale of the Moriscos. This is a clear indication that the authorities believed the rebels attached great importance to the prophecies. The principal royal agent in Granada, Pedro de Deza, devised a plan in which Alonso del Castillo, posing as a fakir, would reinterpret the prophecies to predict Morisco ruin, and call on the rebels to surrender.[150] On 6 February 1570 Castillo wrote a letter in Arabic to this effect; it was taken to various localities in the Alpujarras by a Morisco spy in the service of the duke of Sessa, one of the royal commanders of the campaign. In the letter Castillo wrote 'my powers have allowed me to see that this attempted conquest is a false path, for we have been misled by vain promises' and 'prognostications which are incorrect, for these promise us disaster and not success'. Those who had been encouraged by the year beginning on a Saturday had made the mistake of referring to the Christian solar calendar whereas the prognostication should have been related to 'our lunar calendar'. But there were in any case no good grounds for looking to that day because 'there would be many occasions when the year would begin on a Saturday'. The forecast of one prediction that 'no more of us will die than a single man of low status' was contradicted by the 'more reliable' prediction of a Libyan that 'few of us will be left'. Pointing to other contradictions in the prophecies the pseudo-fakir declared: 'I give little credit to any of them because there is no mention of them in the Koran or in the Law approved by the Caliphs'. No one dared to attack the mighty king of Spain overseas, so 'how can we hope to take what he holds within his own frontiers?'. The best course was to surrender and 'perhaps the king will take pity on us'. The effectiveness of this piece of propaganda led the duke of Sessa to ask Castillo to invent a second letter. According to Castillo, Philip was shown the letters and 'was pleased with the way they had helped to pacify the kingdom'.[151]

Prophecy was the practice of the occult at its most powerful. Seen as a dangerous cause of social unrest, the government of Reformation England had passed an Act (1549–50) prohibiting prophetic utterances, though this failed to prevent Catholic recusants from continuing to see comets and monstrous births as signs of their coming deliverance, or Protestants in Mary Tudor's reign from prophesying the death of their persecutor. In the Iberian peninsula each of the constituent communities entertained Messianic expectations and beliefs that they were God's chosen people. In Philip II's Portugal resistance to Castilian rule was nurtured by the myth of Sebastian's secret survival after Alcázarquivir; comets and new stars were interpreted as signs of his eventual return, and even after their independence (1640) Portuguese, convinced they were the chosen people, continued to prophesy the rise of a Portuguese king who would conquer Jerusalem and the world.[152] In Spain at the time of the Jewish expulsion Rabbi Isaac Abrabanel believed that celestial conjunctions indicated the redemption of Israel. Moriscos looked for natural signs of their liberation by an Islamic conqueror from the East. And Old Christians believed in their divine mission to conquer the world for Catholicism; one clergyman interpreted the comet of 1500 as a sign of the troubles to come from the feigned baptisms of the Moors, and the later comet of 1607 as one of sixteen natural prodigies signifying divine approval for the Morisco expulsion.[153] All were hoping for better days to come.

Horoscopes and prophecies had frequently been associated with the enemies of Philip II and Spanish Catholicism. Yet the crown had good reason to give its strongest support to the observation of the stars for purposes other than astrology.

## Notes

1. *Discurso de la vida del ilustrísimo y reverendisimo Señor Don Martín de Ayala . . . escrito por sí mismo, Autobiografías y memorias*, ed. M. Serrano y Sanz (Madrid, 1905), p. 211.
2. AGS: E 1145/63, Maestre de Campo Don Diego Enríquez to duque de Terranova, 14 April 1576. Sweat from a statue of the Virgin in Aragón was said to have been requested by Philip II 'for his devotion'; Fr. M. de Guadalajara y Xavierr, *Memorable expulsion y justissimo destierro de los Moriscos de España* (Pamplona, 1613), p. 49.
3. BM`Egerton MS. 442 ff.144–6.
4. M. Menéndez Pelayo, *Historia de los heterodoxos españoles*, vol.2 (Madrid, 1880), p. 654.
5. J. Caro Baroja, *Vidas mágicas e Inquisición* (Madrid, 1967).
6. A.Venegas, *Breve declaración de las sentencias y vocablos obscuros que en el libro del Tránsito de la Muerte se hallan, Escritores Misticos Españoles*, ed. M. Mir, vol. 1 (Madrid, 1911), p. 290. First published in 1544, other editions of Venegas' book appeared in 1565 and 1574.
7. B. Porreño, *Dichos y hechos de el Señor Rey Don Phelipe Segundo, el prudente potentissimo, y glorioso monarca de las Españas y de las Indias*. The edition which I read was published in Madrid in 1748 and the anecdotes referred to are on pp. 88, 92, and 99. Although he praises Philip's hostility to astrology Porreño mentions that the damage caused to the Escorial by a thunderbolt in 1577 had been predicted by Micón, a Catalan astrologer (p. 194); that Philip's annexation of Portugal and its empire in 1580 marked the 'sixth monarchy' of the world and that this great event had been signalled by a comet (p. 269); and that Philip's death had been signalled by drought, plague, rising prices, and by three eclipses of the sun and moon (p. 17).
8. A. Rodríguez, 'El pronóstico astrológico que de Felipe II hizo el Doctor Matias Haco', *La Ciudad de Dios*, **97** (1914), 442.
9. Quoted in J. C. Rule and J. J. TePaske, *The Character of Philip II* (Boston, 1963), p. 29.

10. G. de Andrés, 'Los Libros de la Testamentaria de Felipe II (1601)', *Documentos para la Historia del Monasterio de San Lorenzo el Real de El Escorial*, **7** (1964), 397.
11. They include the unfavourable predictions of an Italian astrologer, in the service of the French, that Philip's nativity signified the frustration of his designs and the failure to secure a lasting succession to the throne: Rizza Casa, *Breve tratado di naturale astrologia giudicaria* (Lyons, 1591), p. 95. Matthaeus Delius, a Dane, predicted in 1551 from Philip's nativity that he would not inherit the whole of his father's Empire; L. Thorndike, *A History of Magic and Experimental Science*, vol. 6 (NY, 1941), p. 136. John Dee, Queen Elizabeth's astrologer, had previously prepared horoscopes for Philip and Mary; I. Calder, 'John Dee studied as an English Neoplatonist' (PhD dissertation, London, 1952), vol. 1, p. 311.
12. R. Taylor, 'Architecture and Magic. Considerations on the *Idea* of the Escorial', *Essays in the History of Architecture presented to Rudolf Wittkower*, ed. D. Fraser *et al.* (London, 1967), pp. 81–109.
13. The rich collection of occult works in Herrera's library can be judged from the list appended to Taylor's article, and F. Sánchez Canton, *La librería de Juan de Herrera* (Madrid, 1941).
14. Herrera's letter to Philip on the cranes is reproduced in A. Ruiz de Arcaute, *Juan de Herrera, architecto de Felipe II* (Madrid, 1936), pp. 36–8.
15. G. Kubler, *Building the Escorial* (Princeton, 1982), p. 128f.
16. L. Cabrera de Córdoba, *Historia de Felipe Segundo, rey de España* (Madrid, 1619), p. 1072.
17. BM Egerton MS. 592 ff. 38–48v: 'Discurso astronómico sobre las constituciones generales y significaciones dellas especialmente sobre el cometa que en ix de Noviembre deste presente año a parecido 1577'.
18. B. Barrientos, *Cometarum explicatio atque praedictio* (Salamanca, 1574). He was *regente de gramática* from 1561 to about 1574; it is unclear if after that he held a chair in mathematics.
19. 'Diurnal de Antonio Gracián, Secretario de Felipe II', ed. G. de Andrés, *Documentos para la Historia del Monasterio de San Lorenzo el Real de El Escorial*, **5** (1962), 73.
20. IVDJ 61 (ii)/19.
21. BM Add. MS. 28,360 ff. 1–2: Gesio to the king, 18 February 1578. Although ciphers in the sixteenth century had strong associations with the occult, for example in Trithemius' *Steganographia* where it was a veil for Kabbalistic magic, Gesio's secret code was intended for affairs of state and had no magical connotations.
22. *Ibid.*, ff. 3–4v: Gesio to the king, 25 March 1578, Madrid.
23. 'Discorso di Gio. Batt. Gesio Mathematico di sua M$^{a}$. Catholica sopra la successione del Regno di Portugallo, dirigido a su S.C.R.M.', 18 September 1578, Escorial MS, P.I.20 ff. 1–14.
24. Escorial MS, P.I.20 ff. 151–2: Gesio to Philip II, undated.
25. *Ibid.*, f. 30: same to the same, 2 May 1579.
26. *Ibid.*, f. 36 and f. 38: same to the same, 22 October 1579 and 13 January 1580.
27. *Actas de las Cortes de Castilla*, vol. 3 (Madrid, 1864), p. 407.
28. P. Pierson, *Philip II of Spain* (London, 1975), p.61.
29. *Recopilación de las leyes destos reynos* (Alcalá, 1598), *lib.* viii, *título* iii, *ley* viii. Consistent with this Act, Philip had called for the investigation of Pierola of Navarre whose prophecies from written characters and diagrams were said to have interested even discerning men like Arias Montano; he was subsequently exposed as a trickster and punished; Cabrera de Córdoba. *op. cit.*, p. 1073.
30. For further details, Frances Yates, *The Art of Memory* (London, 1966), chapter 8.
31. Taylor, *op. cit.*, p. 82.
32. R. J. W. Evans, *Rudolf II and His World. A Study in Intellectual History 1576–1612* (Oxford, 1973), p. 218.
33. IVDJ 100/236, Arias Montano to Philip's secretary, Mateo Vázquez, with Philip's holograph indicating his wish to discuss the matter and expedite the acquisition of the books. *Ibid.*, f. 235, letter from a prior, 19 December 1577, informing

Philip that the bishop of Barcelona was organising an immediate search for Lull's works in the region. For the related efforts of Dimas de Miguel, Juan Vileta and Antonio Bellever, and for other information on Philip's patronage of Lull see J. and T. Carreras y Artau, *Historia de la filosofía española* , vol. 2 (Madrid, 1943), p. 257f.

34. Quoted in Carreras y Artau, *op.cit.*, pp 257–8, from the prologue to Seguí's book.
35. Juan Seguí, *Vida y hechos del Admirable Doctor y Martyr Ramon Lull vezino de Mallorca* (Mallorca, 1606), pp. 34 and 37. Missionary zeal was also the motivation behind Pedro Aguilón's Castilian translation of Lull's *Arbre de filosofía desiderat*, presented in 1590 to Isabel, Philip's daughter, in the hope that 'she may intercede with His Majesty for the diffusion of this holy doctrine throughout his realms, and the extension of the faith of Jesus Christ, Lord of the Universe, all over the earth', Escorial MS. d. III.23.
36. Pedro de Guevara, *Arte general para todas las sciencias en dos instrumentos* (Madrid, 1586), from the dedication to Philip II. Guevara was a Lullist priest and tutor to Philip's daughters.
37. For the complicated question of Lull and the Roman Index see H. C. Lea, *A History of the Inquisition of the Middle Ages*, vol. 3 (N.Y., 1888), pp. 585–6.
38. Escorial MS. P.I.20 f. 73–74v, Gesio to Philip II, undated. Parts of Gesio's criticism sound like contemporary Aristotelian opposition to Lullism, seen as a new form of logic to displace Aristotle's scientific method. The hostility of Gesio, a Neapolitan, shows the weakness of Taylor's speculation that Herrera's Lullism may have been brought from Naples, 'one of the great centres of Lullism in Italy,' by Bautista de Toledo who had resided there; Taylor *op.cit.*, p. 82.
39. This, as well as the richness of other occult works, can be judged from P. Guillermo Antolín, 'La librería de Felipe II (datos para su reconstitución)', *La Ciudad de Dios*, **116** (1919), 36–49; 287–300; 477–88; **117** (1919), 207–17; 364–77; **118** (1919), 42–9; 123–137; also G. de Andrés, 'Entrega de la librería real de Felipe II (1576)', *Documentos para la Historia de San Lorenzo el Real de El Escorial*, **7** (1964), 7–233.
40. AGS: E 1503/138, the king to Guzman de Silva, 20 April 1572.
41. *Ibid.*, 1549/44–5, 'Relación de los libros que se han comprado en Venecia por orden de su Mj[d] y de lo que por ellos pago el Embaxador Diego de Guzman de Silva los quales se han embiado a su Mj[d], 12 April 1576, Venice. Most were works of literature, philosophy and theology.
42. *Ibid.*, 1507/134, record of purchase by Guzman de Silva, 17 August 1572.
43. *Ibid.*, 1549/44–5. This purchase included other works on the occult: Porphyry's *De Virtutibus*, Proclus' *Platonic Theology*, Macrobius' commentary on the *Dream of Scipio*; and medical texts.
44. *Ibid.*, 1550/403, 'Memorial de los libros que Christobal de Salazar Secretario de la Embaxada de Venecia embia en ocho caxas a Alicante de Venecia por Abril de 1579'. This consignment consisted predominantly of printed scientific books in Latin and Italian. Several were astronomical, including Reinhold's edition of Peurbach's *Theoricae novae planetarum* (Paris, 1558); others were medical: Celsus, *De Medicina* (Lyons, 1561); Isabella Cortese, *Secreti* (Venice, 1574).
45. 'Relazione di Filippo II Re di Spagna letta in Senato da Michele Suriano nel 1559', *Le Relazioni degli Ambasciatori Veneti al Senato durante il Secolo Decimosesto*, ed. E. Albèri, series I, vol. 3 (Florence, 1853), p. 367.
46. 'Relazione di Filippo II Re di Spagna letta in Senato da Marcantonio da Mula il 23 Settembre 1559', *ibid.*, p. 397.
47. Letter from Le Prévôt Morillon to Cardinal Granvelle, 9–10 September 1581, Mons; *Correspondance du Cardinal de Granvelle (1565–1583),* ed. C. Piot, vol. 8 (Brussels, 1890), p. 404.
48. Escorial MS.P.I.20 f. 73, Gesio to Philip, undated.
49. Extracts from de Hoyo's correspondence with Philip on these experiments, though without any reference to the source of the documents, were published in F. Rodríguez Marín, *Felipe II y la alquimia* (Madrid, 1927), pp. 19–25.

50. IVDJ 61(i)/228v.
51. IVDJ 31M/153. The document is interesting for the terms of a proposed contract with the king, agreed by Bufale and probably showing the results of preliminary negotiation with a royal official: Bufale would be given an immediate payment of 300 *escudos* to cover the cost of experiments and living expenses for five months; he would agree to work under guard; half of the silver prepared would go to Philip, the other half to Bufale. If successful Bufale would serve Philip at his court for eight years and thereafter serve the king in Spain or Italy; in the event of failure he would bear the costs and 'the person and life of the said Marco Antonio put at the disposal of His Majesty'.
52. BM Add. MS. 28,357 f. 41: the king to Juan de Zúñiga, 29 August 1574.
53. IVDJ 61(ii)/261, Antonio Gracián to the king, 10 June 1572. Taylor saw only Gracián's abbreviated record of this correspondence in his diary which, without any mention of the king's scepticism, merely stated that the alchemist was to let Philip know when the work was finished. From this Taylor, *op.cit.*, was led to conclude that Philip had a continuing interest in alchemy. Kubler saw the original document and even reproduces it in a special appendix entitled 'Herrera and the King on Alchemy'; but failing to recognise Gracián's monogram he mistakenly supposed the letter to be from Herrera, and concluded that it was important evidence for Herrera's hostility to alchemy and against the view that Herrera performed occult services for Philip; Kubler, *op.cit.* pp. 129 and 140.
54. Apart from riches and health, some sixteenth-century practitioners, notably Paracelsus, saw in alchemy the higher spiritual aim of man's regeneration, to be achieved by understanding the hierarchies and correspondences in the universe, and manipulating the occult forces of nature. It has recently been suggested that Philip II's interest in alchemy may also have been due to magical inspiration of this type; but the only support adduced for this is the lost book Philip is said to have written 'on the relation between the universe of God and that of man' and his affection for the paintings of Hieronymus Bosch; Javier Ruiz, 'Los Alquimistas de Felipe II', *Historia 16*, **2** (1977), 49–50; the rest of the article contains nothing new.
55. ARP, *sección Administrativa: Cédulas Reales*, vol. 8, f. 443–4 in which Holbeeck's salary is fixed at 300 ducats a year.
56. *Ibid.*, *ibid.*, vol. 9, f. 314, 11 April 1598, appointing Juan de Ausnero distiller of waters and essences, and, as his assistant, Juste de Fraye, 'who at present serves as my distiller in the monastery of San Lorenzo el Real'.
57. IVDJ 99/303, Giovanni Vincenzio Forte to the king, enclosed in a letter of 21 Oct. 1581; ARP, *Cédulas Reales*, vol. 3, f. 256v indicates that Forte was appointed in April 1579. Records at the ARP show that four distillers with the name Forte served at Aranjuez up to 1680.
58. Juan de Almela, a physician of Murcia, gave the first detailed account of the Escorial in his 'Descripción de la Octava Maravilla del Mundo que es la Excelente y Santa Casa de San Lorenzo el Real' of 1594. The final part, which includes the description of the laboratories, has been edited by G. de Andrés in *Documentos para la Historia del Monasterio de San Lorenzo el Real de El Escorial*, **6** (1962), pp.67–9.
59. J. M. López Piñero, *Ciencia y técnica en la sociedad española de los siglos XVI y XVII* (Barcelona, 1979), p. 277.
60. Richard Stanyhurst (1547–1618) was born in Ireland, studied at Oxford and Lincoln's Inn, and while in London in 1578 witnessed experiments to convert copper into silver which 'convinced me of what I had until then thought impossible and led me to apply myself to the practice of this secret science'; 'Toque de Alchimia', BNM MS. 2058, f. 252. He converted to Catholicism and in 1581 settled in the Netherlands, where he continued to practise alchemy and performed cures with chemical remedies, which was brought to Philip's attention. He worked in the Escorial laboratory in 1592–5 and then returned to the Netherlands. He was associated with the regiment of English Catholics of Sir William Stanley, a soldier who converted to Catholicism in the Netherlands and entered Philip

II's service. Stanyhurst conspired with Stanley to restore a Catholic monarchy in England. Although Stanyhurst was not a soldier he was paid as a 'King's Pensioner' attached to Stanley's regiment; a Spanish official described him as 'very loyal to the King; not a soldier but very useful in other ways'; A.J. Loomie, *The Spanish Elizabethans. The English Exiles at the Court of Philip II* (NY 1963), p. 259. There is an entry in the *Dictionary of National Biography* and now C. Lennon, *Richard Stanihurst the Dubliner 1547–1618* (Blackrock, 1981) though this has little on his work at the Escorial, p. 48. See also A.J., Loomie, 'Richard Stanyhurst in Spain: Two Unknown Letters of August 1593', *Huntington Library Quarterly*, **28** (1964–5), 145–155. None of these sources mentions the BNM manuscript.

61. BNM MS.2058, 'Toque de Alchimia', 25 September 1593, San Lorenzo el Real, is the only known alchemical work by him.
62. *Ibid.*, f. 249
63. *Ibid.*, ff. 256–7.
64. *Ibid.*, f. 254.
65. Cabrera de Córdoba, *Historia de Felipe II, op.cit.*, p. 1072.
66. *Lettres de Philippe II à ses filles les Infantes Isabelle et Catherine écrites pendant son voyage en Portugal (1581–1583)*, ed. L. Gachard (Paris, 1884), p. 198. Philip's second wife, Mary Tudor, believed in the curative virtues of gold coins worn during the royal touch, the treatment for scrofula.
67. BM Add. MS. 28,344 f.368, Antonio de Montoya to the king, undated.
68. José de Sigüenza, *Historia de la Orden de San Jerónimo* (Madrid, 1590; 2nd ed., Madrid, 1909), p. 499.
69. From a passage in his *Diálogo de las cosas ocuridas en Roma* (1527) quoted in M. Morreale, 'Comentario de una página de Alfonso de Valdés: el tema de las reliquas', *Revista de Literatura*, **21** (1962), 66. Valdés also regarded the cult of saints as a legacy of paganism: the Christian Santiago had taken the place of Mars; St. Elmo that of Neptune; and the medical dominion of Aesculapius divided so that SS. Cosmo and Damian presided over common complaints, SS. Roque and Sebastian over pestilence, St. Lucia cared for the eyes, St. Polonia the teeth and Sta. Aguda the breasts; *Historia de la Iglesia en España,* ed. R. García-Villoslada, vol.3, part i (Madrid, 1980), p. 374. Similarly Martín de Castañega, a friar, wrote 'in ancient times men had recourse to diabolical superstitions and invocations to Apollo; and in the same way present-day Catholics who lack effective natural medicines seek the intermission of saints', *Tratado de las supersticiones y hechicerías* (Logroño, 1529; reprinted with introduction by A. de Amezúa, Madrid, 1946), p. 76.
70. Accounts of these events are given by Cabrera de Córdoba, *op.cit.*, p. 296, and by a physician who was in attendance, 'Relación de la enfermedad del Príncipe D. Carlos en Alcalá por el Doctor Olivares médico de su cámara', *Codoin*, **15** (Madrid, 1849), 553–74.
71. *Canons and Decrees of the Sacred and Oecumenical Council of Trent*, trans. J. Waterworth (2nd ed., London, 1888), pp. 234–6.
72. J. Corominas, *Diccionario etimológico de la lengua castellana*, vol. 4 (Berne, 1954), p. 803.
73. Examples of this belief in other parts of Europe in the sixteenth and seventeenth centuries are given in Thorndike, *History of Magic*, vol. 6, p. 502. And an early-fourteenth century example of treasure-seeking in Carcassonne by means of a child gazing at oil spread on finger nails is mentioned by H. C. Lea, *A History of the Inquisition in the Middle Ages*, vol. 3 (NY, 1888), p. 437. Aristotle had said that people of 'inferior type', especially the mentally unstable, had a greater sensitivity in divination because their minds, vacant and not preoccupied in thought, were more open to the impressions of the external world (*De divinatione per somnium*, 463b 15–464a 25); the alleged sensitivity of child diviners was explained in exactly the same way by Bartolomé de Las Casas, *Apologética historia*

*sumaria*, probably written in the later 1550s, ed. E. O'Gorman, vol. 1 (Mexico, 1967), pp. 439–40.

74. IVDJ 61(i)/299; Pedro de Hoyo to the king, undated, with Philip's scrawled reply. The rest of the correspondence on which I have based my account survives in IVDJ 61(i)/14; 75; 299; 377; 379; 384; 386; 391 and BM Add. MS.28,350 ff. 315–6, 319, 321v–323v and 325.
75. H. C. Lea, *A History of the Inquisition of Spain*, vol. 4 (NY and London, 1907), p. 187.
76. From the trial fully described in Caro Baroja, *Vidas magicas*, vol. 1, p. 307.
77. T. Sánchez, *Opus morale inpraecepta Decalogi*, posthumous publication (Lyons, 1615), lib. II, c.38, n.37. In Elizabethan England it was also a serious offence to seek subterranean treasure or precious metals by magical means.
78. This widening of authority applies to Castile; in Aragón the Inquisition already had this function.
79. Lea, *Inquisition of Spain, op.cit.*, vol. 4, p. 186.
80. *Ibid.*, vol.4, p. 193.
81. S. Muñoz Calvo, *Inquisición y ciencia en la España moderna* (Madrid, 1977), p. 81.
82. Augustine, *De Civitate Dei*, V, i–x.
83. Aquinas, *Summa Theologica*, IIaIIae, quaestio 92, artic. 1–2 and quaestio 96, artic. 5.
84. Juan de Horozco y Covarruvias, *Tratado de la verdadera y falsa prophecia* (Segovia, 1588). This invalidates the supposed delay in publication of the Bull until 1612, stated by Lea, *Inquisition of Spain*, vol. 4, p.186 and repeated by Caro Baroja, *Vidas mágicas*, vol. 2, p. 180.

84a. M. de Guadalajara y Xavierr, *Memorable expulsion y iustissimo destierro de los Moriscos de España* (Pamplona, 1613), p. 106v.

85. G. Henningsen, 'El "Banco de Datos" del Santo Oficio. Las relaciones de causas de la inquisición española, 1550–1700', *Boletín de la Real Academia de la Historia,* **74** (1977), 547–70; J. Contreras, 'La Inquisición de Aragón: estructura e oposición (1550–1700)', *Estudios de Historia Social,* **1** (1977), 130; J.P. Dedieu, 'Les causes de foi de l'Inquisition de Tolede (1483–1820)', *Mélanges de la Casa de Velázquez,* **14** (1978), 143–171; R. García Cárcel, *Herejía y sociedad en el siglo XVI. La Inquisición en Valencia 1530–1609* (Barcelona, 1980), p. 249. G. Parker, 'Some Recent Work on the Inquisition in Spain and Italy', *J. Modern History,* **54** (1982), 519–32.
86. What happened there is now well known. After the local tribunal had vigorously persecuted 29 alleged witches, the Inquisitor General intervened and sent Alonso de Salazar Frias to visit the area. His enlightened report, which presented the outbreak as an illusion, resulted in a change of policy; after that the Inquisition turned aside from the prosecution of witchcraft. The affair is also notable for the Inquisition's use of scientific tests in gathering evidence: Salazar collected jars of powders and ointments, supposed to be poisons prepared by the witches, and passed them on to physicians and apothecaries, who demonstrated their harmlessness by feeding them to animals. The events are fully described in G. Henningsen, *The Witches Advocate. Basque Witchcraft and the Spanish Inquisition (1609–1614)*, (Nevada, 1980); see also the assessment in H. Kamen, *Inquisition and Society in Spain in the sixteenth and seventeenth centuries* (London, 1985), pp.212–4.
87. The fourteenth-century pope John XXII was supposed to have issued a strong prohibition against alchemical transmutation identifying this with counterfeit. But while Aquinas also thought the sale of alchemical gold or silver was fraud, he added: 'if however alchemists discovered how to make true gold it would not be unlawful to sell it as genuine, because there is nothing wrong in employing natural causes for the production of natural effects', *Summa Theologica*, II IIae, quaestio 77, artic. 2.
88. García Cárcel, *Herejía y sociedad*, p. 260.
89. The case is described in Muñoz Calvo, *op cit.*, p. 44f.
90. The incident which occurred in 1598 was described in M. del Río, *Magicarum*

*Disquisitionum* (Louvain, 1599–1600), vol.2, p.114. The author, a Jesuit, explained that the treatment at Ypres 'expected God to perform a miracle, and to ask God habitually for miracles is to tempt Him'.

91. M. van Durme, *El Cardenal Granvela*, trans. E. Borras Cubells and J. Pérez Ballestar (Barcelona, 1957), p. 289.
92. AGS: E 1529/31, Cardinal Granvelle to Cristóbal de Salazar, 28 January 1583, Madrid.
93. Granvelle to Prieur de Belle-Fontaine, 17 August 1585 in *Correspondance du Cardinal de Granvelle*, ed. C. Piot, vol. 12 (Brussels, 1896), p. 87.
94. References are from the recent translation: *Pedro Ciruelo's A Treatise Reproving all Superstitions and forms of Witchcraft very necessary and useful for All Good Christians zealous for their Salvation*, trans. E.A. Maio and D'Orsay W. Pearson (London and NJ, 1977), p. 58.
95. *Ibid.*, p. 90.
96. *Ibid.*, pp. 172–4
97. M. del Río, *op.cit.*, vol. 1, p. 100.
98. Las Casas, *op.cit.*, p. 500.
99. García Cárcel, *Herejía y sociedad*, pp. 258–9. In May 1586 the Spanish Inquisition in Palermo punished a village priest for conjuring the devil in order to discover treasure; a black mass was celebrated in a cave; Henningsen, 'El "Banco de Datos"', pp. 551–2. The concern of ecclesiastical authorities to eradicate superstitious masses in Spain is briefly mentioned in García-Villoslada, *Historia de la Iglesia en España*, vol. 3, part i, pp. 372 and 378.
100. Archivo Histórico Nacional: *Inquisición, legajo* 4467, no. 4; undated document of the beginning of the seventeenth century.
101. AGS: E 1230/16 and 98, Juan de Vargas Mexia to Philip II, 8 February 1571 and 27 February 1571, Turin.
102. *Ibid.*, 1540/115, Gian Andrea Doria to Francisco de Vera y Aragón, ambassador in Venice, 27 October 1589.
103. *Ibid.*, 1540/335, same to the same, 4 January 1590, Loano.
104. *Ibid.*, 1540/180, duque de Terranova to Francisco de Vera y Aragón, 22 August 1590, Milan. Ten years later the Spanish ambassador in Venice told Philip III that Henry IV was trying to buy 'powders for making gold' which Bragadino had left with the Venetians, and thought this was 'in order to deceive merchants' since the powders could have no value: *ibid.*, K1677, Iñigo de Mendoza to Philip III, 11 March 1600.
105. BNM MS. 5917, 'Astrología Judiçiaria que leío en Sevilla Diego Pérez Messa', 1595.
106. J. Pérez de Moya, *Tratado de cosas de astronomía y cosmographía y philosophía natural* (Alcalá, 1573), p. 112. 'Astrology', 'astronomy' are also distinguished, p. 7.
107. Wellcome Institute MS.99, 'De las Inteligencias de su Govierno Mando y Asistencia a las cosas perecederas y temporales', 1605.
108. *Ibid.*, unpaginated section entitled 'De los números cubos y Rayzes dellos'.
109. *Ibid.*,ff. 27–28v.
110. L. Coçar, *Dialogus veros medicinae fontes indicans* (Valencia, 1589), p. 22v; facsimile with introduction by J.M. López Piñero (Valencia, 1977).
111. F. de Silva y Oliduera, *Discurso en la providencia y curación de secas y carbuncos con contagio* (Granada, 1603), p. 12.
112. *Codoin*, **15** (Madrid, 1849), p. 570.
113. *Ibid.*, p. 573.
114. AGS: E 1520/56, Joan de Arrazola to Guzman de Silva, 12 November 1577, Madrid.
115. *Correspondance du Cardinal de Granvelle 1565–1583,* ed. C. Piot, vol. 6 (Brussels, 1887), p. 592; anonymous communication, November 1577.
116. C. D. Hellman, 'A Bibliography of Tracts and Treatises on the Comet of 1577', *Isis*, **22** (1935), 41–68 lists over one hundred items, but only one from Spain: the *De cometis* of Fernández Raxus, an Aragonese physician, published in Madrid, 1578.

117. Hellman, *The Comet of 1577: Its Place in the History of Astronomy* (NY, 1944), pp. 132–133; Tycho's predictions were not published until the twentieth century.
118. A. W. Lovett, *Philip II and Mateo Vázquez de Leca: The Government of Spain (1572–1592)*, (Geneva, 1977), p. 133 lists books in Vázquez's library.
119. BM Egerton MS. 592, ff. 47–47v., anon., 1577.
120. *Ibid.*, ff. 42–42v.
121. *Ibid.*, f. 44.
122. AGS: E 1547/105. Similar sentiments were later expressed at the beginning of the seventeenth century to justify the expulsion of the Moriscos. Marcos de Guadalajara, a Carmelite, wrote that signs in the heavens were intended by God to be interpreted correctly only by His preferred people and that therefore they were misunderstood by the 'perverse' Moriscos: 'Because of their Catholic faith God has always held the Spanish for His own. They have spread the faith throughout the world and have taken arms against the impious Moor. God has loved them just as he once did Israel, whom he also chose. And to assist them He has made prodigious miracles'. The prodigies included the great conjunction and new star of 1603, and a comet of 1607; *Prodición y destierro de los Moriscos de Castilla hasta el Valle de Ricote* (Pamplona, 1614), pp. 9–24.
123. 'Discurso sobre el cometa que apareció en 13 de Julio de 1596', Escorial MS. L.I.12, reproduced in *La Ciudad de Dios,* **82** (1910), 190–4.
124. *Historia de Felipe II, op.cit.*, p. 962.
125. *Ibid.*, pp. 104–5.
126. 'Discurso á propósito del cometa de 1577 contra la astrología judiciaria', Escorial MS. L.I.12. published in *La Ciudad de Dios*, **82** (1910), 292–8. The anonymous author entertained the common Aristotelian view that comets were hot and dry exhalations, rising from the earth as a result of solar heat, and igniting in the upper regions of the air; but he looked to better explanations in the future.
127. BM Egerton MS. 592, 'Acerca de los cuentos de las Brujas. Discurso de Pedro de Valencia dirigido al ilustrissimo S[r]. Don Bernado de Sandoval y Rojas, Inquisidor general', f. 197v.
128. H. Friedenwald, *The Jews and Medicine*, vol. 2 (Baltimore, 1944), pp. 419–29.
129. J. Ferrandis, *Marfiles y azabaches españoles* (Barcelona, 1928), p. 257.
130. Ciruelo, *op.cit.*, p. 245.
131. Quoted by H. Kamen, *Spain 1469–1714. A Society of Conflict* (London and NY, 1983), p. 180.
132. The work of Joseph ben Ephraim Caro (1488–1575) who was born in Spain and then, like many other Sephardi Jews, settled in the Ottoman Empire.
133. García Cárcel, *Herejía y sociedad*, p. 254. For the circulation of the *Clavicula* in Spain see Caro Baroja, *Vidas mágicas,* vol. 1, p. 135f; for its content *The Key of Solomon the King (Clavicula Salomonis)*, trans. and ed. S. L. Mathers (London, 1909). Garbled Hebrew words were also used for magical medicine in Elizabethan England; K. Thomas, *Religion and the Decline of Magic* (Harmondsworth, 1973), p. 213.
134. Caro Baroja, *op.cit.*, vol. 1, pp. 81 and 91.
135. Caro Baroja, *Los Judíos en la España Moderna y Contemporánea* (2nd ed., Madrid, 1978), vol. 2, p. 324.
136. A. A. Sicroff, *Les Controverses des Statuts de 'Pureté de Sang' en Espagne du XV[e] au XVII[e] Siècle* (Paris, 1960), pp. 114–115.
137. Caro Baroja, *Los Judíos en la España*, vol. 1, p. 189.
138. A. Domínguez Ortiz and B. Vincent, *Historia de los Moriscos* (Madrid, 1978) is a fine recent synthesis.
139. IVDJ 49/94, Francisco de Ibarra to Mateo Vázquez, 5 October 1576, Madrid. The Moor, placed under custody in Ibarra's house, claimed to have made a pound of silver; but Ibarra was not yet convinced.
140. D. Hurtado de Mendoza, *De la Guerra de Granada*, ed. M. Gómez-Moreno (Madrid, 1948), p. 23.

141. A. Gallego and A. Gámir Sandoval, *Los Moriscos del Reino de Granada según el Sínodo de Guadix de 1554* (Granada, 1968), p. 118. The Synod was convened by Martín de Ayala, bishop of Guadix; he himself accepted questionable parts of judicial astrology (see note 1).
142. J. Ribera and M. Asin, *Manuscritos árabes y aljamiados de la Biblioteca de la Junta* (Madrid, 1912), pp. 98–104. In Granada in 1526 the authorities tried to prevent silversmiths making Morisco amulets, and to substitute crucifixes and other Christian images; Gallego and Gámir Sandoval, *op.cit.*, p. 116. This source also states that protective amulets with inscriptions on paper, placed in the foundations and walls of houses, are frequently discovered during the demolition of houses in the region around Granada; *op.cit.*, p. 115. Another scholar confirms the rich magical content of Morisco literature, and finds that the literary output was predominantly concerned with popular religious devotion and hardly at all with natural science; he sees this as evidence that the Moriscos had become introspective in the face of persecution; L. P. Harvey, 'The Literary Culture of the Moriscos' (D. Phil. dissertation, Oxford, 1958).
143. J. Fournel-Guerin, 'La pharmacopée morisque et l'exercice de la médecine dans la communauté morisque aragonaise (1540–1620)', *Revue d'histoire maghrebienne*, **6** (1979), 53–62.
144. M. García-Arenal., *Inquisición y Moriscos. Los procesos del Tribunal de Cuenca* (2nd ed., Madrid, 1983), p. 110.
145. L. García Ballester, *Historia social de la medicina en la España de los siglos XIII al XVI. La minoría musulmana y morisca* (Madrid, 1976), p. 77f.
146. *Ibid.*, p. 149.
147. L. del Marmol Carvajal, *Historia del rebelión y castigo de los Moriscos del reino de Granada* (Madrid, 1858), p. 174.
148. Castillo was a rare example of a university-trained Morisco physician. As a translator from Arabic to Castilian he was employed in succession by the *Chancillería* (high court of justice) in Granada, by the tribunal of Granada, and then by Philip II. His other translations include inscriptions on the Alhambra; correspondence between Philip II and the Sultan of Morocco; an inscribed Turkish banner captured at Lepanto; and the spurious lead tablets, forged texts dug up in Granada, which predicted universal victory for Christianity through the assistance of God's chosen people – sincere Morisco Christians like Castillo. Philip also employed Castillo to catalogue Arabic MS. at the Escorial; D. Cabanelas Rodríguez, *El Morisco granadino Alonso del Castillo* (Granada, 1965). In the last year of his reign Philip brought to the Escorial Diego de Urrea, professor of Arabic at Alcalá, to teach the language to some of the monks. The motives were mixed: to provide linguists for affairs of state; to facilitate 'the extirpation of the Mohammedan sect'; and to benefit from the wisdom stored in the Escorial's great collection of Arabic books (predominantly medical and pharmaceutical); N.Morata, 'Un catalogo de los fondos árabes primitivos de El Escorial', *Al-Andalus*, **2** (1934), 87–181 contains Urrea's account of his mission and an inventory of the Arabic works.
149. Alonso del Castillo, 'Sumario e recopilación de todo lo romanceado por mí. . .', *Memorial Histórico Español*, **3** (1852), p. 67f. and 80f.
150. *Ibid.*, pp. 13–14 describes the plan, and pp. 15–20 contains the text of the letter in Castilian.
151. *Ibid.*, p. 9.
152. BM Add. MS 20,934, 'Algunas consideraciones y reflecciones para consuelo de los Fieles de la Catholica Iglesia Romana sobre las profecías y vatocinios', 1711; ff. 265–8v describe current Portuguese aspirations. Another manuscript records a predicted invasion of Andalucía by Turks and Arabs, and the subsequent emergence of 'a secret king' who, coming forth with a friar and 'a noble of the house of Toledo', repels the infidel and proceeds to the capture of Jerusalem; it

is not clear if the author was a Sebastianist or a Spaniard: MS. Collection Edouard Favre, Bibliothèque publique et universitaire (Geneva), vol. 66, ff. 139–140, 'Relación sumaria de un pronóstico de Andrés de Narbaíz', 1580.

153. M. de Guadalajara y Xavierr, *Memorable expulsion* (Pamplona, 1613), p. 48f., and his *Prodición y destierro de los Moriscos* (Pamplona, 1614), pp. 16v–24.

# 2

## Cosmography and the crown

Astrology had attracted the attention of the crown on more than one occasion. There had been prosecutions by the Inquisition to curb the excesses of judicial astrology; and during the Granada uprising royal agents had exploited Morisco anxiety over signs in the heavens. But the study of the stars was much more than an occasional interest for the crown; it was indispensable for the accurate mapping of the earth and charting of the seas, matters which had become of great importance to the efficient government of Spain and her empire. The king and certain of his councils were therefore constantly concerned with the promotion of cosmography, the science which described the world, both the heavens and earth, and which offered the precious key to sure navigation.

Cosmography had been the most comprehensive of the sciences ever since Pliny's encyclopaedic treatment of the cosmos, and this very broad study persisted in sixteenth-century Spain : the royal cosmographers appointed by the crown were expected to supply information on astronomy, geography, navigation and sometimes natural history as well. Cosmographical treatises conventionally began with a brief description of the arrangement of the planets and the distribution of land and sea on the earth.[1] The universe which they depicted was in the prevailing tradition of Aristotle and Ptolemy: the earth at the centre and at rest; a sublunar region of change and decay, constructed from the four elements, and inferior to the quintessential and eternal celestial region which lay beyond the moon. These fundamental assumptions about the world, commonly taught at universities in Spain as in the rest of Europe, and integrated with Christian doctrine, were challenged on two notable occasions in Philip's reign. How did the government react?

## *1 Criticism of Aristotelian cosmology and its repercussions*

In November 1572 Jerónimo Muñoz, professor of Hebrew and of mathematics (including astronomy) at the university of Valencia, learned from local shepherds that a bright new body had appeared in the heavens. This was the nova near the constellation of Cassiopeia which was exciting astronomers and divines throughout Europe. In Madrid the king's courtiers were wondering whether it was a star or comet, and when Philip heard that Muñoz was making observations he sent an urgent request for their immediate publication.[2] Perhaps Philip had residual anxieties about the supposed connection of comets with the death of kings; what is clear is that the publication of Muñoz's book was a royal initiative and that the text shocked several who read it.

Muñoz had soon decided that the new body was a comet from its fiery colour and apparent size, larger than any star; and he predicted that though it signified war and the ruin of harvests, the king was not threatened. So far there was nothing to shock opinion. But he then presented uncompromising and unpalatable conclusions from careful observations of the position of the shining object which demonstrated that it was situated beyond the moon: in agreement with Tycho Brahe and other observers, he found that there was no parallax, which meant that the body was located at a great distance from the earth. Brahe believed this to be a new star, and his account which implied a falsification of the supposed unchanging character of the supralunar region is usually presented by historians of science as the first great observational assault on Aristotelian cosmology. In fact Muñoz's book preceded Brahe's and went far beyond him or any other author of the time in an attack on tradition which was uncompromising and of outstanding clarity and force. He blamed Aristotle for misleading mankind into believing that the heavens were unchanging and made of a special material; somewhere in the Milky Way a process of ignition had generated this comet, and the formation of a fire in the heavens meant that comets were no different from the earth in their elementary composition; the quintessence was a fiction. Muñoz said it was 'because God has given me an independent mind' that he had been able to recognise the falsity of the generally accepted Aristotelian image of the world.[3] But the hostile reception of his ideas soon caused him to retire into seclusion, expressing regret for the money he had spent in publishing his book and his determination to keep his thoughts in future to himself: 'I do not wish to stir up the hornets any more'. The hornets were 'numerous theologians, philosophers and courtiers' who had 'showered me with insults'.[4] The theological opposition had come from individual conservative ecclesiastics and not from the Inquisition; nevertheless Muñoz was angry with royal authority – the king had instigated the publication and Muñoz alone had been left to suffer the consequences.

Traditional cosmology was again brought into question in Spain in 1584, when Diego de Zúñiga, a monk and recently professor of Holy Scripture at the university of Osuna, accepted the Copernican hypothesis of a planetary earth. He had been working on a commentary to the book of Job and after struggling with the verse (9:6) which states that God 'shaketh the earth out of her place' he concluded that it was comprehensible only by assuming that the earth was in motion. And he added that Copernicus' heliocentric system gave a much better account of planetary motions than the accepted geocentric astronomy of Ptolemy's *Almagest.* Nor was he deterred by the various biblical references to a moving sun; he thought these could be accommodated to Copernicanism.[5]

In later decades it was precisely this confident reinterpretation of biblical passages in defence of Copernicanism which led to Galileo's arrest by the Inquisition in Rome. But in Philip II's Spain Zúñiga was not prosecuted by the authorities. The king had shown his determination to control thought in the universities with his instructions (October 1558) to the rector of Salamanca to investigate the circulation of theologically suspect books, and his commissioning (1560–1) of Bishop Diego de Covarrubias to visit the same university.[6] But here the motive had been the prevention of Protestantism. Copernicus had not been a Protestant and was not seen by the king or the Spanish Inquisition as a threat to Catholic orthodoxy. The king himself owned a copy of Copernicus' *De Revolutionibus.*[7] And at Salamanca in 1561 Copernican astronomy was even allowed a place in the curriculum, the only university in Europe to do so. The professor of astrology was authorised to lecture on Ptolemy's *Almagest* or Copernicus according to the wishes of the students. But no evidence has been found that a course on Copernicanism was ever given, probably reflecting the conservatism of the academics.[8]

All the available evidence suggests that there was no risk to supporters of Copernicus during Philip II's reign. It is true that Zúñiga later retracted (1597) and returned to the stationary earth of Aristotle and Ptolemy. But there is nothing to indicate that this was the result of the Inquisition's intervention, and seems instead to have been due to the impossibility of reconciling the earth's diurnal motion with Aristotelian physics. It was not until after his death that Zúñiga's commentary on Job was condemned, and in Rome, not Spain, along with Copernicus' work in the famous edict of 1616. But in Spain the statutes of Salamanca of 1625 continued to permit the teaching of Copernicus' doctrines if there was student interest for it. Apart from Zúñiga – and then only in his earlier work – it is difficult to find convinced Copernicans in Spain; as in other parts of Europe very few were committed to the physical truth of the heliocentric system, and more used it mathematically as a calculating device for the purposes of navigation.

## 2 *Longitude and politics*

Cosmographers were appointed by both Charles V and Philip II to resolve continuing uncertainties on Spain's territorial possessions. Ever since the fifteenth-century voyages of discovery, Spain and Portugal had come into conflict in their competition for conquest. And both monarchies had come to regard the scientists as authoritative arbiters of these disputes.

By the treaty of Tordesillas (1494) it had been agreed to fix a line of demarcation 370 leagues west of the Cape Verde islands; all discoveries or conquests of non-Christian lands to the east of this line would fall to Portugal, those to the west of it would go to Castile. The settlement was much less precise than it appeared. From which of the Cape Verde islands was the count to be made? It was in the interests of Portugal to refer to Santo Antão, the most westerly of the isles; while Spain wanted the demarcation to be referred to Boa Vista some 150 miles further to the east. There were differences also on whether the distance was to be measured along the latitude of Cape Verde or along the equator, and disagreement on the lengths of a league. According to one of Philip II's cosmographers there were no less than 54 different ways to interpret the Tordesillas line.[9] But these difficulties were petty compared with the determination of longitude, essential for any fixing of positions to the east or west of a specified meridian such as the Tordesillas line. The scientifically and technologically complicated problem of finding longitude could not be solved until the eighteenth century, despite the incentive of a large monetary prize from Philip III (1598). Spain might claim all lands reached by sailing up to 180 ° west of the Tordesillas line, but there was no available means of telling whether or not they were intruding into Portugal's half of the globe. Consequently for most of the sixteenth century the two monarchies continued to squabble and, in remote regions of the earth, to war over their territorial rights. For the support of their respective cases they relied on the opinions of their cosmographers.

When after Magellan's voyage Sebastián del Cano returned to Spain from the Moluccas (1522) with a valuable cargo of cloves, the fierce protests of John III of Portugal – the Portuguese were already in Ternate – had led to negotiations at Badajoz (1524). For this purpose which he regarded as 'of such importance to our service', Charles V had summoned cosmographers to join lawyers in discussions with a similar body of Portuguese experts to press his case for the ownership of the Spice Islands. In the preliminary manoeuvering objections had forced each side to replace their delegates: Charles V's choice of the astronomer Simon de Alcazaba was unacceptable to the Portuguese on the grounds that he was of Portuguese origin and a turncoat. Soon the usual disagreements on measuring the demarcation limit became apparent; but the Portuguese request for new astronomical observations was dismissed by the Spanish as too time-consuming and seen as an attempt to delay a decision in the face of a weak case.[10] The conference had failed to reach an agreement, and the subsequent fighting which broke out in the Moluccas came to an end only through Charles' decision to abandon his claims to the isles by selling them to the Portuguese monarch for 350 000 ducats. By this treaty of Zaragoza (1529) another arbitrary dividing line was created, this time in the Pacific, 297½ leagues to the east of the Moluccas, and it was agreed that no Spanish vessel could enter to the west of it. This was a second line that was impossible to fix without an accurate knowledge of longitude.

Around the beginning of Philip II's reign, Alonso de Santa Cruz, a renowned royal cosmographer, was asked to provide the crown with a survey of the various methods for finding longitude which had been proposed since antiquity. His account,[11] dedicated to the king, assessed the relative merits of the different methods for navigational purposes, and is interesting for its recognition of the obstacles preventing a solution. The determination of longitude from observations of eclipses had first been suggested by the ancient Greeks; the difference in the time of a lunar eclipse seen from two different locations would indicate their difference in longitude. As Santa Cruz rightly commented, this was not of much use for the constant needs of navigators because eclipses were such infrequent phenomena. There were other difficulties with this method. He realised that it was essential to be sure that the same phase of the lengthy process of an eclipse was being timed in the two locations concerned – an error here of just a few minutes in the timing would introduce large errors in longitude. Above all an accurate timekeeper was indispensable; but he said clocks usually gained or lost as much as half an hour a day; they were quite useless for finding longitude.[12] Accurate clocks were also essential for another, more recently suggested method of finding longitude at sea.[13] If a clock set for the port of departure was taken on board ship, the vessel's distance in longitude from the port could be found during its voyage from the difference between the clock and the local time indicated by observation of the sun. But for this to be practicable the clock would have to be accurate to a few seconds a day, otherwise the accumulating error would entail large errors in longitude. Santa Cruz thought that this method was potentially the best, but recognised that there appeared to be insuperable difficulties to maintaining exact timekeepers at sea: the ship's motion affected the running of the clock; so did hot and cold weather, moisture and rust. These were not resolved until Harrison's marine chronometer of the early eighteenth century. An indication of how wide of the mark estimates of longitude could be is seen in Santa Cruz's conclusion that the difference in longitude between Mexico City and Genoa was 217 ° 30 ′, about double the actual figure.[14] It led one reader of his book to complain that by this gross exaggeration of the distance to Mexico City, Santa Cruz had reduced the claims Spain could make to lands further to the west in the Pacific and so had 'given the Portuguese something to talk about';[15] in particular he was concerned about Philip's rights to the Moluccas, still a live issue.

The cosmographers were well aware that existing astronomical data was unreliable and resorted to other sources of information for their estimates of longitude. Calculations were made from Ptolemy's data to assess distances between places in the Old World, and from accounts of Portuguese navigators for regions to the east of the Iberian peninsula. To determine which lands lay in the half of the globe allotted to them, the Spanish deduced what they could from their pilots' logs and counted 180 ° of longitude westwards from the peninsula across the empty spaces of the Pacific. This drew on mariners' estimates of the direction and speed of ships, the technique of 'dead-reckoning', based on observation of the marine compass and crude measurements of ship speed; no astronomical observations were involved. In the hands of a navigator as skilful as Columbus dead-reckoning could be effective; but these calculations of position were ruined whenever the course of a ship had been affected by strong winds or ocean currents. And in the vast expanse of the Pacific there was no guarantee that the pilot had steered the most direct route or that the occasional measurements of a ship's speed could be applied to the entire voyage. Returning Spanish pilots gave the distance from Navidad (New Spain) to Cebu (Philippines) very differently from 1550 leagues to 2260, an underestimate of the actual distance (2400 leagues) by up to 850 leagues (some 3000 nautical miles). According to Sancho Gutiérrez, a royal cosmographer, most of these pilots had overestimated and, judging from the duration of the voyage, he supposed the true distance was nearer 1750 leagues.[16]

With such levels of uncertainty the door was open for cosmographers to supply propaganda in support of extravagant royal claims to territorial possession. The Portuguese seem to have produced two sets of cartographical information in the sixteenth century: one set, supposedly portraying truer distances, intended for themselves and carefully guarded; the other set, distorted to serve their territorial claims, was released as propaganda. The Spanish regularly accused them of deliberate falsification. Santa Cruz said that when he was in Portugal in 1545 the famous cosmographer Pedro Nuñez was instructing all mapmakers to contract the extent of sea between Cape Comorin and Malacca in accordance with the wishes of João de Castro, the viceroy of Portuguese India, who was determined to ensure that the Moluccas remained Portuguese.[17] And López de Velasco, cosmographer of Philip's Council of the Indies, complained that the Portuguese were continuing to adjust maps for their own purposes and to make unjustifiable claims to the Moluccas in the east, and the Castilian settlements of Asunción and Río de la Plata[18] in the west; he preferred to base geography on celestial observations even though exactness was unattainable. But the Spanish could also play the same game. Sancho Gutiérrez advised the king to shorten the distances on maps between the Canaries and the Lesser Antilles, between Cuba and Vera Cruz, and between the east and west coasts of New Spain; if the Atlantic was contracted Spain's 180 ° of longitude would secure more of the Pacific and the demarcation line would fall on Malacca, leaving the Moluccas well within Spanish territory.[16]

Philip's views on these matters can be discerned from his policy on exploration in the Pacific. When he wrote to Luis de Velasco, viceroy of New Spain, in September 1559 to arrange for the preparation of an expedition to the isles of the Pacific, he made it perfectly clear that 'under no circumstances' were the ships to enter the Moluccas; Charles V's concession must be observed. Instead the voyage should be to neighbouring isles: to the Philippines (already named after Philip in 1545) and other isles within Spain's demarcation. The king's intention was 'to procure spice'. European demand for nutmeg and mace from the Moluccas had increased and Philip was hoping for similar valuable supplies from his own territories. On the viceroy's recommendation he approached Andrés de Urdaneta, an experienced navigator and skilled cosmographer, who had sailed to the Moluccas in the 1520s, residing there for eight years; returning to the west, he had then entered the order of St Augustine and settled in New Spain. The monk was persuaded to resume his navigational interests but his reply must have surprised the king. He warned that the king's plans were open to objection because, as he correctly indicated, the Philippines were situated to the west of the Moluccas and consequently within the terms of the Zaragoza treaty which excluded Spanish vessels west of a line drawn 297½ leagues east of the Moluccas. A voyage to secure trade or settlements in the Philippines would therefore be an infringement of Portuguese rights. But he believed the king had legitimate grounds for entering the isles: to rescue Spaniards reported to be held there by infidel.

With these reservations Urdaneta agreed to take part in the expedition. He was soon taking an active part in the preparations, advising on the most suitable months for departure and the best route. He sent requests to the king for mariners' charts, compasses, astrolabes, cross-staves and hour-glasses, the common instruments of navigation which he said were unobtainable in New Spain; and he asked the king to send the best pilots so that an accurate estimate could be made by dead-reckoning of where Spain's 180 ° of longitude really extended to.

Four ships with 380 men left Navidad in November 1564 under the command of Miguel López de Legazpi. It has been alleged that only after Urdaneta was aboard and the ships had put to sea were sealed instructions from the *audiencia* of New Spain opened, ordering Legazpi, contrary to Urdaneta's wishes, to take possession of the Philippines.[19] Urdaneta was in charge of the navigation. His outstanding qualities were evident when he alone identified isles which had come into view as the Marianas, while other pilots mistook them for the Philippines; and above all in his magnificent return voyage across the Pacific, one of the feats of sixteenth-century navigation, which discovered the elusive favourable winds at the high latitude of 42 °N. After five months at sea without touching land he arrived at Acapulco in October 1565, establishing the route to be taken by the Manila galleons over the next two centuries. He had used his knowledge of winds and currents but his astronomical observations during the voyage are not recorded.

Legazpi had in the meantime captured Cebu and would soon establish the Spanish settlement at Manila. Philip was uneasy about the colonisation and received little comfort from the views expressed by Urdaneta and other cosmographers summoned to Madrid in October 1566 to give their scientific verdicts before the royal Council of the Indies. Each was asked to reply to two questions.[20] Were the Philippine isles conceded to the Portuguese monarch by Charles V's sale of the Moluccas in 1529? Were the Moluccas and Philippines Castilian domains as defined by the treaty of Tordesillas? The cosmographers were unanimous on both points. Urdaneta repeated what he had told the king before Legazpi's expedition had set out: the king was not entitled to enter or trade with the

Philippines until the 350 000 ducats paid by the Portuguese monarch in 1529 was returned, because those isles were 'well within' the zone reserved for the Portuguese by the treaty of Zaragoza. And the same response was given by five royal cosmographers: Pedro de Medina, Francisco Faleiro, Sancho Gutiérrez, Jerónimo de Chaves, and Alonso de Santa Cruz. This must have displeased the Council or the king because some months later Santa Cruz and Pedro de Medina returned to sign statements retracting their previous responses. Santa Cruz said unconvincingly that he had no intention of declaring that the Philippines had been conceded to the Portuguese or that Spanish vessels could not land there; that was for lawyers to judge from the text of the treaty. He had only meant to indicate the geographical position of the isles. Pedro de Medina similarly stated that it was no part of a cosmographer's profession to interpret treaties; nevertheless he added that the Portuguese monarch had no absolute sovereignty over the Moluccas and neighbouring isles but only held them as a pawnbroker (plate 1).

On the second question the cosmographers judged that irrespective of the sale of the Moluccas in 1529, which was subject to redemption, the isles were situated within Castile's half of the globe indicated by the Tordesillas line of demarcation. And that by the same treaty the Philippines, Borneo, Japan and the coast of China all fell to the monarch of Castile. The evidence the cosmographers adduced for this came from various sources. Gutiérrez studied the charts of Spanish pilots who had sailed to the Philippines and he used Ptolemy's data to calculate the longitude from the Canaries to the Ganges. Chaves also calculated from Ptolemy's statements of longitude and supplemented this source with a determination of the difference of longitude between Toledo and Mexico City from the observation of two lunar eclipses in New Spain in the 1540s (the figure deduced was 102 °; the actual difference of longitude is 95 ° 7 ′). Urdaneta's remarks on his own calculation are interesting for their reference to Fr Martín de Rada of Pamplona, one of the five Augustines who had gone with Legazpi to the Philippines as missionaries. Urdaneta had asked Rada to take with him an astronomical instrument 'of middling size to determine the longitude of wherever we should arrive, which happened to be the isle of Cebu'. From Rada's celestial observations Urdaneta used Copernican tables to calculate that Cebu was 215 °15 ′ west of Toledo (actually 232 °); and subtracting 43 °8 ′, which he assumed

Plate 1 Portrait of Pedro de Medina, cosmographer royal. He holds a marine astrolabe, part of the standard equipment of Spanish navigators. The design of this instrument became the subject of an official inquiry which sought to provide pilots with a device which was accurate without being unwieldy. In the background are Seville and the Guadalquivir, the centre of Spain's commerce with the Indies. (Museo Naval, Madrid).

separated Toledo from the Tordesillas line, he concluded that Cebu was 172 °7 ′ of longitude west of the Tordesillas line, showing that this isle (and all land 7 ° 53 ′ further to the west of it, to complete the 180 °), by the treaty of 1484, belonged to Castile. According to Urdaneta, Rada was 'a good mathematician, astrologer and cosmographer', and his observations were reliable 'because I know he would only speak the truth of what he had found,' probably an allusion to the contemporary practice of falsifying longitudes.

It is clear from all this that the Moluccas question had not been settled by the sale of 1529. Their status and that of many other lands in the Far East remained contentious. Gesio, Philip's Neapolitan cosmographer, went so far as to advise the king that he would be perfectly within his rights to seize the Moluccas 'without returning one *real*' to the Portuguese monarch. He argued this from the judgment of canon lawyers that any contract yielding a profit of over half the price paid was usurious and invalid; the profits enjoyed by the Portuguese monarch from the sale of the Moluccas spices since 1529 must have been so great that even if Charles V had 'pawned those isles for 5 million in gold, that would have been little'.[21]

Gesio had made a special study of the demarcation dispute and in the 1570s he repeatedly warned the king of the disastrous consequences of Portuguese usurpations in the east. He often advised action that was rash, but his cosmographical knowledge won him some influence with the Council of the Indies. In 1569 he had gone to the embassy in Lisbon, a nerve centre of demarcation affairs. The ambassador Juan de Borja was gathering all the intelligence he could on Portuguese cartography, collecting maps, documents, and narratives of Portuguese voyages. He put Gesio to work on these and reported that after three years' concentrated study the cosmographer had 'become very experienced and resolute on the subject'. And now the ambassador was sending Gesio to Madrid with numerous maps, and geographical and historical texts. Urging the king to arrange for Gesio to interpret these documents to the Council of the Indies, Borja stressed that 'the documents which Gesio carries are important because they clearly show how false maps are made in this kingdom and that the Moluccas are many degrees within out demarcation'.[22] The papers included several early Portuguese charts and descriptions of sea routes to India which, according to Gesio, gave stronger support to Castilian territorial claims than did the maps currently prepared in Lisbon; an early Portuguese text put the Cape of Good Hope 4 ° further to the east than the Portuguese were now prepared to admit. The king asked López de Velasco, cosmographer of the Council of the Indies, to make a preliminary inspection of the documents. Velasco was impressed and asked that Gesio be thanked for his labour; he thought that some of the texts looked as though they could be valuable for the demarcation dispute, recommended that they be translated, and advised that those which proved to be important should be deposited in the archives of the Council of the Indies.

The documents had been collected through the assistance of a Portuguese cosmographer, Luis Jorge de Barbuda, who, Gesio told the king, also 'revealed many secrets to us' and gave information on the falsification of maps in Lisbon because 'he has always wanted to serve Your Majesty'. The value placed on such experts can be judged from the events which followed instructions to Borja, now returning to Madrid, to bring Barbuda with him. King Sebastian learned about the plot and Barbuda was arrested on the border at Olivença, taken back to Lisbon in shackles, and kept in prison for two years awaiting trial; he was eventually released after giving certain assurances. A few years later a second attempt to bring him to Spain was more successful. In 1579 Gesio went to Lisbon and smuggled Barbuda out 'with the utmost secrecy and subterfuge' and, safely back in Madrid, informed the king that 'he is now here and I have him in my house'.[23] Barbuda was employed by Philip with a salary of 150 ducats a year, working on the correction of false nautical charts issued by the Portuguese.

It was on Gesio's advice that the Council of the Indies withheld licences to publish geographical treatises judged prejudicial to Castile's claims. His guiding principle was that since astronomical observations were not yet able to establish the demarcation of territory, the best course was to argue from the earliest charts and histories of the Portuguese, because until they had reached the Moluccas in the early sixteenth century they had prepared their charts and histories of voyages 'according to the truth' which 'they would not have done had they foreseen the subsequent demarcation disputes'. Therefore Spain should draw heavily on these documents allowing the Portuguese 'to condemn themselves by their own evidence which declared in Castile's favour'.[24] But the Spanish must not fall into the same trap; so far he said the Portuguese had been unable to find anything in the charts and histories of the Castilians to support their case. Consequently Gesio advised the king that the greatest vigilance was needed in inspecting Spanish texts which touched on geography to ensure before publication that they contained nothing prejudicial which the Portuguese could make use of. When he heard that the bishop of Manila had issued a bull in which the Philippines were said to be situated '2000 leagues or 114 ° of longitude from the coast of New Spain,' actually in keeping with the estimates of several Spanish cosmographers but not with Gesio's gross under-estimate of '1100 or 1200 leagues,' he expressed his utter dismay in a letter to the king. The bishop's bull 'had done more damage than any previous publication,' sanctioning the loss of 51 ° of longitude to the Portuguese, giving them 'Japan and all of the east up to New Guinea'. He urged the king to take immediate steps to arrange for the destruction of the bull in Rome because 'if it falls into the hands of the Portuguese they will always use it for longitude and demarcation, since it gives them more encouragement than any other history'. But Gesio feared it was already too late; the Portuguese ambassador in Rome was 'very diligent and would probably already have got hold of a translation of the bull'.[25]

Gesio attached considerable importance to Spain's continued possession of the Philippines. Luzon, the largest of the islands, was 'the key to the entire east' and should be regarded as highly as Flanders or Italy'. This was because of its strategic geographical position, surrounded by China, Siam, Malacca, the Moluccas and Japan. He assured the king that in the future Luzon would become 'one of the principal and richest places in the world'.[26] But that was not all. Luzon should become the base 'for the enterprise of Japan and, of much greater importance, of China'[27], nothing less than a scheme for the Spanish conquest of the entire Far East. Gesio thought Philip would have no difficulty in subjugating Japan, thereby winning many converts to Christianity and swelling Castile's treasury through its trade and allegedly rich silver mines. Gesio believed the process of conversion would be facilitated by sending Hieronymite monks whose rituals bore a superficial resemblance to those of the natives. Cosmology had an important part in these plans for conversion; Gesio wanted the missionaries who were sent to be conversant with natural philosophy, because by familiarising the Japanese with the idea of the Aristotelian unmoved mover and how all phenomena followed from this, the idea of God and the essentials of Christian faith could be gradually introduced. Philip never undertook the enterprise. But Gesio's plan was not dismissed as absurd; Benito López de Gamboa, a councillor of the Indies who concentrated on demarcation affairs, was sufficiently moved by the proposal to advise the secret dispatch of missionaries from the Philippines to Japan and the issuing of orders for a ship to reconnoitre the Japanese islands.[28]

When the Portuguese were granted permission by the pope to create the see of Macao (1575), Gesio saw his plans for Japan in ruins and the act as a violation of the demarcation. Macao, according to his calculations, was situated 15 ° beyond Portuguese territorial limits. The Portuguese had established a trading settlement there around 1557 using it to control the trade between China and Japan; and now, Gesio complained, the establishment of the see gave them the opportunity to send Jesuit missionaries to China, Japan and other Castilian territories in the east. When Juan de Zúñiga, Philip's ambassador in Rome, first heard of this he had attached no importance to it because 'I did not know where Macao was situated nor in whose demarcation the province of China fell'; but he later acknowledged that it was a serious matter and that Spain had to achieve the repeal of the see because – and here he accurately portrayed the usual practice of Portuguese monarchs in colonisation – the sending of missionaries to China would be accompanied by the landing of troops there. Gesio thought Zúñiga's ignorance inexcusable; if the ambassador did not know the approximate longitude of Macao he should at least have known that it was in the East Indies and that its elevation to a see incorporating China and Japan within its diocese was prejudicial to Spain's interests.[29] In fact the whole of the Far East was much more ambiguous with respect to longitude and demarcation limits than Gesio supposed.

Some of the tensions between the two monarchies were soon lessened, not by any accurate and accepted cosmographical measurement of longitude but by the union of crowns which followed Philip's annexation of Portugal. The king and the councillors of the Indies had no longer to be concerned with news like Gesio's recent communication of intelligence from Lisbon that a Portuguese attack on the Philippines was imminent. But the tensions did not disappear. If the Portuguese were allowed to continue their trade from Macao to Japan, there was some competition from the Spanish. And the Portuguese Jesuits in Japan and China were joined by Spanish missionaries, although it was not until Philip III's reign that the crown took steps to achieve what Gesio had wanted: papal repeal of the Portuguese monopoly. The Moluccas were now controlled by the Portuguese subjects of Philip II and remained Iberian until their capture by the Dutch in the early seventeenth century.

## *3 Royal projects to survey Spain and the Indies*

The crown also undertook comprehensive surveys of Spain and the Indies which were unprecedented in their scope. Around 1566 the king engaged Pedro de Esquivel, professor of mathematics at Alcalá, to prepare a detailed geography of Spain including information on 'all places, rivers and mountains, large and small'.[30] Felipe de Guevara, who was well-informed about the project and whose son Diego was to continue it, thought:

> it was without exaggeration the most careful and accurate description ever to be undertaken for any province since the creation of the world. . . because all descriptions which have so far been given by Ptolemy and the rest have been based on the accounts of provincials and information taken from others; whereas Your Majesty has ordered that not a handsbreadth of land is to pass without inspection by the author, using his own eyes and hands, so that the truth of everything can be established, in so far as his mathematical instruments allow.[31]

Accurate surveys of Lorraine and the Rhineland had been accomplished earlier in the century; but nothing as ambitious had ever been attempted for an area as large as Spain. A Spanish historian has argued persuasively that the conception of Esquivel's survey was a sign of maturing 'political outlook' in sixteenth-century Spain, and refers to the view of Furió Ceriol, a royal councillor and the author of a work on politics dedicated to the king, that 'to understand the politics of a land it is essential to know the details of towns and territories'.[32]

Esquivel's geodetic survey employed astronomical observations, terrestrial measurements and triangulation techniques. He used long measuring chains and invented large wooden instruments. But the survey of Spain was never completed. First Esquivel, and then his successor Diego de Guevara, whom he had instructed, died within a short interval, leaving the king anxious to retrieve all instruments and papers relating to the project and to try to involve Juan de Herrera in its prosecution. For reasons that are not clear the project was not continued. A detailed map of the Peninsula dating from this time is conserved in the Escorial (Castilian MS. K.I.1) and while it cannot be ruled out that it incorporates some of Esquivel's data, its provenance remains unknown.

More was achieved from the king's commissioning (1570) of the Portuguese cosmographer Francisco Domínguez to undertake a full geographical survey of New Spain. New light on his activities is provided by a legal document[33] which originated from his attempt to secure royal recognition of his services and the pay long due to him. Domínguez initiated proceedings before the *audiencia* of New Spain in which he and several witnesses gave testimony under oath that he had peformed various tasks for the crown without receiving any payment. After his arrival in 1571 he had worked for five years on the description of New Spain, travelling from Yucatan to New Galicia through rugged terrain in a testing climate and at his own expense. First drafts of a detailed description of these provinces, including accounts of the customs of the Indians, had been shipped to Spain. Still awaiting his salary, he proceeded to compile a 'general table' listing all the 'bishoprics, jurisdictions, justices and convents' of New Spain. And from his experience of surveying he was employed by the viceroys to settle out of court several disputes over land boundaries. This was another type of demarcation dispute which cosmography was called upon to resolve. There had been a long-standing quarrel between the archbishop of Mexico and the bishop of Michoacan over the rights to the tithes of Querétaro, which lay between the two dioceses. Through his cosmographical skills Domínguez was able to provide the key to a settlement.[34]

When in 1579 heavy rains caused Lake Texcoco to rise alarmingly and threaten to inundate Mexico City, Domínguez was called in by the viceroy, Martín Enríquez, to inspect the site of a proposed emergency drainage channel suggested by the municipal authorities. He persuaded the viceroy not to adopt this remedy because of the magnitude of the task – the digging of a deep channel 10 leagues long – so saving an estimated expenditure of 200 000 ducats and avoiding heavy manual labour which 'would finish off the few Indians who are left here'. The waters subsided; but Domínguez was annoyed that he had 'received not even a word of thanks'.

For the next viceroy, the conde de Coruña, he compiled geographical descriptions of the newly discovered lands of New Mexico; these too were sent to the king. At the request of viceroys he observed lunar and solar eclipses, preparing a book on 'the true longitudes and latitudes of New Spain'; and he communicated with correspondents in China, sending them instructions for observing eclipses and then incorporating their findings and other information in a geographical 'table' of China. He had been continuously occupied with the construction of cosmographical instruments: clocks, cross-staves, marine compasses, maps, and regiments of the sun. And he had instructed pilots on 'better and safer navigation to Peru and the Philippines'.

For all these services to the crown over more than twenty years he had received no salary. Now he was asking the king for money and the grandiose title of 'cosmographer and chronicler of New Spain, the Philippines, China and adjoining regions, Peru and all other lands of the Indies'. The evidence heard by the viceroy, Gaspar de Zúñiga y Acevedo, and other officials of the *audiencia* convinced them that Domínguez was a dedicated and skilful cosmographer who deserved to be rewarded. That was the recommendation sent to the king in May 1594 and for a second time in June 1597. It was not until September 1598 that the king, now on his deathbed, sent a reply: a request for further information, asking the viceroy if the office Domínguez sought was really necessary and if so what his salary should be. Before this was taken any further Domínguez became ill and died, naming as his sole beneficiary the hospital of San Hipolito (Mexico City), with instructions that its administrators continue the dispute with the crown and take possession of the money due to him; officials of the hospital were still arguing for this in the 1610s.

The variety of Domínguez's work in New Spain was typical of what was expected from a royal cosmographer. When Jaime Juan of Valencia was sent by the king to the Indies in 1583 it was primarily to determine the latitude of localities in New Spain and the Philippines and, from lunar eclipses, their longitudes. But he was also instructed to observe and record the variation of the ship's compass during his voyage out and to measure the altitude of the sun wherever the ship landed; to watch pilots and mariners 'to see how their practices could be improved' and to teach them how to use navigational instruments; to make maps 'wherever he goes' and record descriptions 'of anything notable'.[35] This mission, estimated to require 6–8 years, was cut short by Juan's death from fever after his arrival in the Philippines.

Even broader concerns became the responsibility of the first cosmographers appointed to the Council of the Indies. The *Real y Supremo Consejo de las Indias* was the consultative council on whose advice the king relied for the government of the Indies. Created in 1524 it supervised the administration of all affairs concerning the Indies except for finance (transferred at the beginning of Philip's reign to the *Consejo de Hacienda*) and the maintenance of religious orthodoxy which was the responsibility of the Inquisition. But the Council's lack of knowledge about the Indies shocked Juan de Ovando during the inspection which he made on the king's orders in 1569. He advised reform and with Philip's support the Council assumed a new scientific complexion, adding to its existing membership of half a dozen lawyers a 'cosmographer-chronicler' who was clearly introduced to remedy the lack of information. The revised code of regulations for the government of the Indies, issued in September 1571, specified the considerable duties of the new office. The cosmographer-chronicler would be required to compile 'all the routes and navigations' from Spain to the Indies, and from one part of the Indies to another, basing this on reports received from pilots and mariners. He had to construct 'cosmographical tables' indicating the longitude and latitude of places in the Indies and the distances in leagues between them according to the information reaching the Council. He was instructed to supervise the determination of longitude from lunar eclipses and send to royal officials in the Indies the times when they were to be observed and a description of the techniques to be employed. And he was to collect all the information he could on the natural history of 'herbs, plants, animals, birds, fish and other notable things'. As if this were not enough, he was also asked to write a general history 'as accurate as possible' of the 'events, antiquities and customs' of the Indies.[36]

The office would later be divided with the separate appointments of cosmographer and chronicler; but in 1571–88 Juan López de Velasco was appointed to do all of these things with an annual salary of 150 000 *maravedís*. Some of the information which was called for was no different in type from what the first *conquistadores* had been asked to provide. But this enterprise was distinguished from all previous inquiries by its scale and organization, by the persistence of the crown, revealed in repeated royal demands for information requested but not received; and by the form of the inquiry – an elaborate standard questionnaire printed with precise instructions on how to complete it. This had been perfected by López de Velasco from questionnaires on the Indies drawn up by Ovando.

By May 1576 printed copies of 49 questions were ready for distribution throughout the Indies. In a royal cedula addressed to the governors of provinces and sent with the questionnaire, Philip explained that the purpose was 'the achievement of good government,' and that after 'frequent discussion' with the Council of the Indies it had been agreed that this could be secured by 'a general description of the entire condition of our Indies, as precise and as accurate as possible'. He gave orders for the distribution of an appropriate number of copies within the various jurisdictional units and for the return of completed forms 'as quickly as possible' to the Council. The king's interest in the project is apparent here; apart from his desire for strong and efficient royal control of the Indies, the systematic collection of information was a method of procedure which appealed to him – on another occasion his passion for relics led him to suggest a complete survey of holy relics conserved and revered in the peninsula.

From the viceroys the questionnaires went down the governmental hierarchy until they reached the smallest municipal authorities or, 'where these do not exist,' the clergy. The responses were to be completed by local government officials or 'intelligent persons' selected by them. Every question was to be considered 'responding to each by its number and leaving a blank if no answer can be given'. The most important final instruction asked for 'brevity and clarity in all, affirming what is certain and indicating what is dubious'.[37]

To answer the questions it was sometimes necessary to consult Indians, which required knowledge of local languages. This would be provided by men like Felipe Guacra Páucar, an Indian who had resided in Spain, and who served as an interpreter in the completion of the questionnaire for the province of Xauxa (Peru); in Quacoman (New Spain) the questions were put by the *alcalde mayor*, Baltazar Davila Quiñones who had learned the language of the Mexican Indians in his district.

Some of the questions concerned physical geography, asking whether the terrain was flat or mountainous, requesting a list of rivers and lakes, mountains and volcanoes, and information on climate. Question 6 asked 'if there was anyone in the locality who knows how to take the elevation of the pole star' and if so to record latitudes determined by this method. Illustrated maps or sketches of towns and ports were requested; also the distances in leagues between *pueblos* and of these from centres of government, with descriptions of the roads which connected them. The questionnaire shows a conspicuous interest in matters relating to human economic geography. Was the region healthy, and if not what was the cause? What were the prevalent diseases and what herbal remedies were used by the Indians? Was there a shortage of water and what were the prospects for future irrigation works? What use was made of the timber and fruit of indigenous trees, and what new uses could be found? How successful had been the transplantation of trees from Spain, and the introduction of wheat, barley, vines, olives and livestock from the peninsula? In addition to the flora and fauna, questions on metals, precious stones and salt deposits completed an inquiry into local natural history which was framed from a strongly utilitarian point of view.

Not surprisingly several of the questions reflected a pronounced interest in navigation; nine of the 49 questions directly concerned this. The crown wanted precise information on whether seas were stormy or calm, the times of tides, the location of reefs and shallows, the prevalence of ship-eating marine worms and other hazards to navigation. Full information on ports was expected: the depth of their water, the state of the sea-bed at their entrances, their approximate dimensions to show the sizes of vessels which they might receive, and their facilities for 'firewood, water and refreshment'. The remaining questions concerned the origin of towns; the numbers and way of life of the Indians, and the tribute they paid to the crown; and details of religious and charitable institutions.

The important matter of longitude was not included in the questionnaire. Probably because of the difficulties associated with its determination it was given separate treatment. López de Velasco later circulated advice on how to observe lunar eclipses including detailed instructions on the construction of simple instruments: the preparation of flat surfaces on which were placed simple dials made from paper circles with iron or wooden gnomons positioned at their centres; the points on the dials where the shadows of the sun at midday and at sunset fell, and similarly for the shadow of the moon at its setting and during the beginning and ending of the eclipse were each to be marked.[38] López de Velasco observed the time of the same eclipse in Madrid to provide the remaining data needed for a determination of longitude.

The replies to the questionnaires, or '*relaciones*' as they were called, began to reach the Council of the Indies in 1577–8. These were from Venezuela; returns from New Spain arrived in 1579–80 and from Peru in 1582–8.[39] The returns show that questions were often left unanswered, certain of them repeatedly. Question 6 on latitude proved particularly troublesome. From Trujillo (Peru) the *corregidor* Francisco de Acuña could reply to this that 'no response could be given and there is no one who knows how to take the elevation' of the pole star. And the answer which came from La Paz stated that 'there was no pilot to be found in this city nor any equipment for taking the elevation of the pole star'; but the latitude 'is thought to be 23 ° or 24 °S ''. This was grossly inaccurate, the actual latitude being 16 °30 ′. The much more demanding determination of longitude seems to have received little attention despite the importance attached to it and the attempt to simplify the procedure. Francisco Domínguez and Jaime Juan, royal cosmographers in Mexico City, observed lunar eclipses in accordance with the instructions; but elsewhere little seems to have been done. In Panama a lunar eclipse was observed on 15 July 1581 by Cepeda, president of the *audiencia* of Tierra Firme, using the recommended dials; but clouds and rain had hindered the observations and to measure time he had nothing better than an hour-glass turned every half an hour. It was calculated that Panama was situated '49¼ ° in longitude from the meridian of the Canary Isles'[40] (the true value is 61 °– 66 °, depending on which of the Canary Isles is taken as the reference point).

The *relaciones* were never combined to form the envisaged 'general description of the Indies', but they supplied the crown with a considerable body of detailed information, including scores of illustrated local maps, far exceeding what had previously been available. Today they provide the historian with one of the richest sources on sixteenth-century America. The Indies survey stimulated an analogous crown project for the Peninsula, where similar printed questionnaires were circulated in 1575; but *relaciones* exist only for the region of New Castile. Here the officially stated intention was to provide 'a description and history of the towns of Spain for the honour and ennoblement of these kingdoms'. There was no mention of the need to secure 'good government', probably to avoid arousing suspicions amongst the suffering Castilians that the purpose of the questionnaire was to levy still more taxes.

## 4 *Provisions for the perfection of navigation*

The duties of most of the royal cosmographers were concerned chiefly with the sea. The advantages to the crown of efficient navigation were underlined in the prologue of Pedro de Medina's *Arte de navegar* (1545) which was addressed to the future Philip II: navigation had been the means of securing the monarchy's enormous territorial expansion in the New World; a treasure of gold and silver was repeatedly being brought back to Spain in ships; and, of still greater value than this, navigation permitted the monarchy to propagate the true faith in parts of the globe separated from the peninsula by thousands of leagues of sea, so facilitating the conversion of large numbers of idolaters to Catholicism. But Medina thought there was no art more dangerous than navigation; the lives of all who sailed and precious cargoes were constantly threatened by tempests, hurricanes and the turbulent currents of the Indies route which frequently caused ships to go off course: 'I have often seen pilots return from our Indies after experiencing great danger, having been even on the point of death, and yet soon after their arrival they forget it like a dream and then they prepare to return as if it was a pleasure. This is not out of greed but the result of divine will; because if the dangers were recalled no one would navigate'.[41] A pilot encountered no less danger than a soldier on a battlefield; like the cavalry 'he has arms to defend himself, which are his instruments (compass, chart, astrolabe and cross-staff), accurate and carefully adjusted; and his horse is the ship'.[42] But ships and lives continued to be lost because pilots had not learned enough of the art, and that was why he had written his treatise. Navigation based upon 'arithmetic, geometry and astrology' had acquired such certainty that it was possible for a pilot to indicate the location of 'a rock 200 leagues from land even though there is only the sky and the water' and no landmark, plotting the dangerous object on a chart so that 'he and others will know how to avoid that place'.[41] Solomon had thought sailing out of sight of land 'difficult but if he were alive now how much more so would it seem to him' to navigate over such extensive and dangerous seas![43]

The accuracy of nautical instruments and the adequacy of pilots' training were recurring matters of concern for Philip II and his advisers, as they had been since the time of the Catholic Monarchs. Occasionally the king appointed individual royal cosmographers to ensure the success of important nautical enterprises. Urdaneta's participation in the voyage to the Philippines is one example of this; another is the attachment of Álvarez de Toledo to Pedro Menéndez de Avila's armada of galleons which patrolled the Atlantic shipping routes; there is record of the eight wooden and brass navigational instruments with which Álvarez was supplied by the Council of the Indies with instructions to use them to determine the armada's positions in latitude and longitude and to follow the variation of its marine compasses.[44] The king was fully aware of the need for pilots: 'they are the principal instruments of navigation and it is of the utmost importance that there is always a large number of them available'.[45]

To supply well-trained pilots the king created new teaching centres. In 1575 the crown was seeking a 'professor of the art of pilotage and navigation' to teach pilots on the north coast; a course of instruction lasting four months followed by an examination was envisaged. Andrés de Poza was appointed as an itinerant professor to teach at various ports along the coast; his salary of 200 ducats a year was to come from the sale of licences to export timber from Galicia to Andalucía.[46] And in the Academy of Mathematics which Philip established in Madrid in December 1582 navigation was prominent in the curriculum. João Lavanha, a Portuguese *converso* and distinguished cartographer – he had earlier taught mathematics to King Sebastian – was brought over 'to organise in our court matters relating to cosmography and geography' and was soon giving public lectures on the use of nautical instruments and the construction of charts. He was later joined by Juan Cedillo Díaz, who similarly wrote a navigational treatise based on his teaching at the Academy. The king's hopes that the curriculum of the Academy would be imitated in urban schools throughout Castile were dashed by the refusal of the towns to subsidise the scheme.[47]

But it was the older institution of the *Casa de la Contratación* (House of Trade) at Seville which continued to function as the principal centre for navigational teaching in the Peninsula. Created by the Catholic Monarchs in 1503 as an institution for organising trade with the Indies (and Barbary), its officials (a treasurer, a factor and a notary), appointed by the crown, communicated with government personnel in the Indies to learn what merchandise was in demand from the Peninsula and what was to be shipped from the Indies. These officials also sought captains, prepared fleets, supervised freight and recorded the imports of American bullion which went to warehouses of the *Casa*. Soon the *Casa* had acquired additional functions. The crown authorised it to act as a law court (from 1511) with sole jurisdiction over disputes relating to the *carrera* (the Indies trade), and in cases involving infringement of its regulations; the *Casa* even had a prison. The *Casa* also became a scientific institution with the establishment of its first scientific office, that of pilot-major (1508) with duties to examine and license pilots, and to supervise the production of navigational charts and instruments. At about the same time the crown had given instructions for the maintenance of a *padrón*, a standard reference chart of the Indies prepared and kept at the *Casa*; it was the model from which all permitted mariners' charts were produced and was subject to continual correction from the reports of returning pilots. A second scientific official had been introduced in 1523: the cosmographer-major who was now responsible for supervising the manufacture of all charts and nautical instruments.

The *Casa* was governed by the crown through the Council of the Indies which directed its activities and from time to time made visits of inspection. During Philip's second regency it became clear that the *Casa* was in need of reform; abusive practices were interfering with the scientific and commercial functions of the institution. In August 1552 Philip intervened with orders that henceforth the pilot-major was forbidden to receive gifts, to make navigational instruments or to teach.[48] This can only mean that pilots were passing their examinations through bribes; that the pilot-major had been profiting from his powers to authorise the adoption of navigational instruments; and that undesirable consequences had resulted from his examination of student pilots he himself had been teaching. In December 1552 Philip arranged for the separation of teaching and examining at the *Casa* by creating a third scientific office, 'a chair of the art of navigation and cosmography' with a salary of 30 000 *maravedís*. The preamble of the royal cedula stated that 'we are informed that masters and pilots who sail to the Indies are untrained and without the requisite skill in navigation, causing many difficulties and the loss of ships and lives'. To rectify this the new chair was to be established and a one-year course of study introduced which all intending pilots and masters would have to attend and pass before they received licence to operate on the Indies route.

The new appointment went to Jerónimo de Chaves, the son of Alonso de Chaves, a cosmographer at the *Casa* since 1524 and the recent successor to Sebastian Cabot as pilot-major. This was not sheer nepotism; Jerónimo was an able cosmographer[49] who had written an impressive commentary (1545) to Sacrobosco's *Sphere*, and his *Chronographia* (1548), a work on time, chronology and the calendar, went through numerous editions. His nomination was made by Philip after consultation with the officials of the *Casa*. Philip's cedula specified what the new professor was to teach. Sacrobasco's *Sphere*, for centuries the traditional introduction to Ptolemaic astronomy, was to be studied; the basic skill of setting a ship's course by pricking off marine charts with the points of dividers; the use of 'regiments of the sun' (rules for determining latitude from the elevation of the sun above the horizon, measured by a marine astrolabe, and from the consultation of tables giving the sun's daily position with respect to the celestial equator[50]); the determination of latitude by observation of the pole star;[51] the markings of compasses to determine magnetic variation in any location; the use of diurnal and nocturnal clocks; and the times of tides were all to be taught. The greatest importance was attached to an understanding of 'the theory and practice' of the compass, astrolabe, quadrant and cross-staff, 'so that errors in them can be detected'.

Much of this was not new; the emperor's regulations of 1527 had already required intending pilots to show an understanding of these instruments – at the examination specimen instruments were actually brought in by the pilot-major to test the knowledge of the candidates;[52] and on each occasion a licensed pilot departed for the Indies he was required to present his instruments for inspection to ensure they were fit to use. Nor was the union of theory with practice an innovation; that had been a guiding principle since the early years of the *Casa*: every licensed pilot was to be trained in the theory of navigation and examined 'irrespective of their experience'.[53] What was new in Philip's regulations was the greater concern for accuracy, and that continued to be shown in the later enactments of his reign. But the impressive curriculum was soon weakened by the reduction of the period of study first from one year to three months (legislation of Juana, princess regent, June 1555) because 'masters and pilots are poor', then to two months, and finally to two months 'including feast days' (legislation of Philip II, 1567–8).

This was the state of the organisation of pilot training at the *Casa* which so aroused the admiration of Stephen Borough, the English navigator who visited it in 1558, that he pressed for similar provision in England, advising the creation of a pilot-major.[54] But subsequent legislation makes it clear that the *Casa* was far from a model of perfection; in spite of its recent reform it was not functioning satisfactorily either in its scientific teaching or in its control, intended to be total, over maps and instruments used for navigation to the Indies. The king had to intervene twice in 1564–5 to deal with inefficiencies in the inspection of instruments.[55] The regulations required the pilot-major and the cosmographers to meet once a week to inspect, approve and stamp all navigational instruments issued for voyages to the Indies; but there were too many instruments for the staff to deal with – they seem to have been given too little time in the hectic period of preparations for the departure of the *flotas* – and on the day of their meeting the examination of astrolabes was sometimes prevented by overcast skies; checks on marine compasses were generally neglected. As a result *flotas* had left Seville for their long voyage to the Indies 'without carrying the necessary instruments which is very dangerous for navigation'. The king therefore instructed the scientific personnel of the *Casa* to meet twice a week for the inspection of instruments and suggested that those found to be defective should be 'destroyed or stamped with two R's as a sign of rejection'. After consultation with the officials of the *Casa* it was agreed that henceforth inaccurate astrolabes would be 'broken up and refounded', defective cross-staves and compasses destroyed if they could not be put right, and charts which were too faulty to be corrected cut into pieces 'so that they cannot be used again'. It was also agreed that more rigorous controls were necessary to ensure that those who gave the stamp of approval to instruments were in no way open to the temptations of personal profit; the pilot-major and cosmographers who had the powers of approval or rejection must have no involvement in the manufacture or sale of instruments. The seals of approval[56] were to be kept in the chest of the *Casa*, its two keys restricted to the pilot-major and 'an expert pilot' who assisted in the inspection of instruments.

In January 1568 and again in March 1575 Philip received complaints from the Council of the Indies that the *padrón*, the master chart of the Indies routes kept at the *Casa*, was not being adequately corrected because of the failure of pilots to comply with regulations requiring them to keep daily records of their voyages, noting shallows, reefs, currents, latitudes, and other important information, and to present these on their return to the cosmographers of the *Casa* so that new data could be identified and the *padrón* brought up to date. The king sought to rectify this by instructing the *Casa* to impose heavier fines (an increase from 2 to 4 *reales* in 1568; 'suitable penalties' in 1575) on offending pilots.

The training and licensing of pilots by the *Casa* was also criticised. In 1586 the king instructed its officials to rectify the examination of pilots which 'is not always done with the requisite care, and so inept persons are licensed causing damage and losses'.[57] Anxiety over the examination was again apparent in 1596 when the king sent to Seville García de Céspedes, a royal cosmographer who was not attached to the *Casa*, to assist in the work of that institution. His declared duties, indicating inadequacies at Seville since these were the normal tasks of the *Casa*, included the correction of the *padrón* for which 'the pilot-major was to make all papers available to you'; the production of six other maps of greater scale covering all parts of the world; the design of an improved standard astrolabe, cross-staff and marine compass; and an investigation of whether these common nautical instruments should be replaced by others. Privately he was also instructed to report to the Council of the Indies on the adequacy of the lectures given to pilots and on their examination, because 'so many lives, ships and so much wealth are in their hands that it is most desirable that they have the requisite knowledge and experience and that they are examined with the appropriate rigour'. And reflecting the crown's uncertainty over the organisation of the *Casa*, García de Céspedes was also asked to report on whether it was best for the office of the examining pilot-major to be kept separate from the professor of cosmography who did the teaching, or if the same person should do both.[58] In fact the same person was already doing both: Rodrigo Zamorano, professor of cosmography and cosmographer in charge of maps and instruments, had also taken over as pilot-major, replacing the ageing Alonso de Chaves. Zamorano denied that this presented him with an easy way to pass his own students, indicating that he could be outvoted by the examining board and that in universities teachers were not prohibited from examining their students for degrees.[59] The king ordered García de Céspedes to act as pilot-major until the matter was settled; two years later Zamorano was reappointed.

How did the pilots respond to their training? Historians so far have said little about pilots as a group. In the sixteenth century derogatory remarks were often made about them by cosmographers and others. Gian Andrea Doria, Philip's captain-general of the sea, described Spanish pilots as '*ruines*' (despicable)[60]; and the cosmographer Martín Cortes, who had experience of teaching them in Cádiz, commented on their 'ignorance' and revealed the frustration of a disappointed teacher: 'the regret is not so much that they do not know, but that they do not want to know'.[61] But new information from documents in the Archive of the Indies at Seville puts the pilots in a more favourable light. Towards the end of the reign, the officials of the *Casa* decided to distribute a printed sheet to pilots requesting their views on three specific points. Were the maps of the *Casa* in need of correction? Should the marine astrolabe be enlarged? Was it desirable to modify the design of the marine compass to facilitate the observation of variation? The replies – more than 30 of which survive in the form of brief responses written over the printed circular – also record opinions on wider issues.[62] Several of the replies display complacency; others show a willingness to learn, a readiness to innovate, and a concern for accuracy. Diego Sánchez thought the lectures given at the *Casa* for pilots were 'good and necessary' and made the perceptive remark that when different pilots recorded conflicting measurements with the cross-staff 'it is not the fault of the instrument but of those who use it'.[63] Jerónimo Martín wanted every pilot to test any new modification of the compass, taking the altered form with the conventional instrument on voyages because 'it is important that we experiment with much care to see which of them is better'. Another pilot asked for the compass to be made larger 'for greater division and precision'. And several recommended the construction of larger astrolabes for the same reason, to provide a more finely divided dial 'allowing readings of half and quarter degrees'. But the marine astrolabe in use – it was a highly simplified relative of the medieval astrolabe, retaining no more than the alidade with its pinnule-sights moving over a scale graduated in degrees – was metal and weighed as much as 10 pounds; and now in the interests of greater accuracy, some pilots were asking for this to be increased to 13 pounds. Although it could be suspended on deck, Spanish pilots seem to have preferred to hold it in the hand when taking their measurements of the sun's elevation; the difficulties of holding the instrument steady on board a rolling ship to allow a careful measurement can be imagined. Jerónimo Martín therefore advised a compromise: the astrolabe must not be so large that it becomes 'too heavy and tires the hand' nor so small that its divisions prevent an accurate measurement. The call for greater accuracy was also evident in the request of some pilots for maps of larger scale to register important local features like 'the shallows called "*isla de arenas*," 5 or 6 leagues off Tierra Firme'. The pilots differed considerably in their opinion of the charts supplied by the *Casa*. Some expressed general satisfaction on the grounds that 'our predecessors have done well with them' and that although some judged them faulty because ships had been lost 'this has not been due to errors on the charts but to currents or to fortuitous events which would not have occurred if the ships had sailed in summer'.[64] Others notified corrections with model precision, like the pilot who indicated that 'off the coast of Campeche near Sisal, at a latitude of 21½ °, there is a shallow lying NW–SE with little more than one fathom of water, and I think it is about 10 or 12 leagues from land'.

A few were sharply critical: 'the entire coast of the South Sea from Peru to New Spain has to be amended on the *padrón*'; 'the entire Newfoundland coast needs greatly to be corrected'.

How many attended the course for pilots at the *Casa*? Some indication of numbers is given by Zamorano's statement that in the twelve years since his appointment as professor of cosmography (1575–87) he had taught 'over 280 of them'.[59] There is abundant evidence that there were never enough pilots to meet Spain's needs. The king had instructed (1572 and 1587) every galleon of the armada to carry two examined and licensed pilots: a chief pilot and an assistant in case one became ill or died during a voyage, leaving the ship without a guide; this doubling requirement could not be met. Shortages were also due to the fact that pilots were specialists trained for particular routes; one pilot would describe himself as 'examined for Santo Domingo, Havana and New Spain,' another for Tierra Firme, and neither was able to switch from one route to the other without additional experience. This even applied to the coasts of the peninsula. In 1584 Francisco Duarte, factor of the *Casa*, was instructed to send out two caravels from Seville to meet the *flota* returning from Tierra Firme at the Azores and transmit orders to put in to Lisbon because of storms off the coast of Andalucía; he replied that a dozen Portuguese pilots should go with the caravels 'because it is quite likely that of all those on the *flota* there is not one [pilot] who will be able to do this navigation because they have not been examined and do not know' the coasts of Galicia and Portugal.[65]

To remedy these shortages Philip reluctantly conceded the right to employ foreign pilots. Regulations prohibiting this had been in force since 1515 when the Portuguese 'however learned in navigation' were excluded. Now there were other undesirables. In 1567 all pilots and masters who were supplied with charts by the *Casa* had to swear, under grave penalties, 'not to give, lend, sell or in any way show a nautical chart to foreigners because they could then go to the Indies and become corsairs'.[66] But when supply ships for the armada against England were being prepared in July 1588 neither Castilian not Portuguese pilots could be found with experience of the coasts of Asturias, Vizcaya, and France; French pilots were accordingly recommended. Philip relaxed the regulations in 1589 allowing the pilot-major to examine 'foreign Catholics, especially Italians and excluding only the English'. And again in April 1595 the king permitted the use of foreign pilots except French, English or rebel Dutch, but only 'out of necessity and for the present' because it was 'very undesirable to open the door to them'.

Poor pay seems to have been one cause of Spain's shortage. In April 1590 a search for pilots in Guipúzcoa for the galleons of Castile brought the response that there were few to be found and that 'none of them were willing to go' on the pay offered of 13 ducats a month.[67] In desperation one naval commander wrote to the king: 'although some shortages in the Armada can be overlooked, that of pilots cannot because without this type of person misfortune is bound to occur'.[68] It was not the only deficiency the king had to contend with in his preparations for war.

## Notes

1. M. Cortés, *Breve compendio de la sphera y de la arte de navegar*, the best known work of this type, was first published in Seville in 1551; there were numerous later editions and translations into English.
2. The king had earlier employed Muñoz to survey rivers for an irrigation project in the Murcia region. On that occasion Muñoz, using an astrolabe, made an accurate determination of the latitude of Murcia, recording 37 ° 57 'N., just 2 ' from the true value; Francisco Cascales, *Discursos históricos de Murcia y su reyno* (Murcia, 1621), p. 262v.
3. J. Muñoz, *Libro del nuevo cometa* (Valencia, 1573). This has recently been published in facsimile together with other astronomical texts by Muñoz and with an introduction by V. Navarro Brotóns (Valencia, 1981); the accompanying English translation is imperfect. Muñoz's observations were praised by Tycho Brahe in *Astronomiae instauratae progymnasmata* (Uraniborg, 1602), p. 567.
4. Muñoz to Bartholomew Reisacher, 13 April 1574, Valencia; reproduced in Navarro Brotóns, *op. cit.*, p. 101. The same letter indicates that the king's request had led to a hastily published book marred by errors. Reisacher, who taught mathematics in Vienna, agreed with Muñoz on the absence of parallax but denied that the body was new. He interpreted it as a star which had suddenly become visible through some clearance of the region between it and the earth, so saving the unchanging Aristotelian heavens; this was also the view of the Spanish physician, Francisco Valles.
5. The extracts from Zúñiga's works which refer to Copernicanism are conveniently brought together in *Materiales para la historia de las ciencias en España: s. XVI–XVII*, ed. J. López Piñero, V. Navarro Brotóns and E. Portela Marco (Valencia, 1976).
6. M. Fernández Alvarez, *Copérnico y su huella en la Salamanca del Barroco* (Salamanca, 1974).
7. Philip purchased a copy in Spain in 1545; P. Guillermo Antolín, 'La librería de Felipe II (datos para su reconstitución)', *La Ciudad de Dios*, **116** (1919), 42f. The king's considerable interest in astronomy is shown by his personal collection of instruments. In the Museo Arqueológico (Madrid) there is a fine astrolabe, 59 cm in diameter, made for Philip by Gualterus Arsenius of Louvain in 1566; and in the Museo Naval (Madrid) an elaborate folding case of navigational instruments, constructed by Tobias Volckammer, comprising an astrolabe, quadrant,

compass, calendars, engraved maps of the northern and southern hemispheres, and the longitude and latitude of principal European cities. There was also talk of establishing an observatory at the Escorial with instruments made by the royal cosmographer, García de Céspedes; but that came to nothing.

8. V. Navarro Brotóns, 'Contribución á la historia del copernicanismo en España,' *Cuadernos Hispano-Americanos*, **283** (1974), 3–24.
9. AGI:P 259/79, f.8v, G-B. Gesio, report to the *Consejo de las Indias* on a geographical work entitled 'Sumario de las Indias,' 11 April 1580, Madrid.
10. MN:FN XVI/42/399–407, views of the Spanish delegation at the Badajoz conference, 1524.
11. A. de Santa Cruz, *Libro de las longitudines y manera que hasta agora se ha tenido en el arte de navegar*. I have consulted the edition published in 1921, Seville. Santa Cruz had been a cosmographer in Charles V's court and he continued to serve Philip II in this capacity. He prepared numerous maps. When Philip asked him for a map showing the Turkish incursions into Persia, he was able to correct his first version with the help of Michelle Mossora, a Venetian who had come to Madrid to offer his services to Philip – he knew Turkish, Greek and Albanian – in the event of an expedition to the Levant. Mossora's 'experience of Turkish wars' allowed him to indicate the situation of various places on Santa Cruz's map; AGS: E 1498/254, Michelle Mossora to the king, 3 May 1567, Venice.
12. Because of the rapidity of the earth's rotation each point of the surface of the globe rotates 360 ° in 24 hours, or as Santa Cruz indicated, a difference of one hour in local time between two places was equivalent to a difference in longitude of 15 °. The contemporary difficulties of timing lunar eclipses were apparent in the observations made in Mexico City in November 1584 by the Valencian astronomer Jaime Juan. When the moon was first visible that evening the eclipse had already begun; it lasted five hours. Juan therefore had to time the moment when the eclipse ended and in the end recorded three quite different times: 7.31 pm from 'a clock with wheels which indicated hours and minutes, and which was set as well as could be done'; 7.22 pm and 8 seconds, a time deduced from the observed position of a fixed star; and 7.20 pm and 20 seconds, calculated from the observed elevation of the moon; AGI:M.y P. Mexico, 34, 'Observación del eclypse lunar que aconteció el año de 1584 a 17 días de noviembre en Mexico ciudad. . .por la qual se contó la diferentia de la longitud y la distancia que ay desde la ciudad de Mexico y la de Sevilla'.
13. Gemma Frisius of Louvain had proposed this in 1522.
14. The calculation given in his *Libro de las longitudines*, pp. 62–3 was based on his observations of the moon in Mexico City and ephemerides calculated by Pedro Pitato for Verona, which 'is almost in the same meridian as Genoa'. The huge error was due to the inadequacy of contemporary lunar tables; these were based on observations with instruments which were too coarse to establish the complicated motion of the moon.
15. BM Add. MS.33,983 ff.267–267v, Pedro Juan de Lastanosa to unnamed correspondent, 13 January 1568.
16. MN:FN XVII/24/249–249v, formal statement of opinion by Sancho Gutiérrez to the *Consejo de las Indias*, 8 October 1566, Madrid.
17. BM Add. MS. 17, 625 f.99, statement of Santa Cruz before the *Consejo de las Indias*, 8 October 1566, Madrid.
18. J. López de Velasco, *Geografía y descripción de las Indias*, ed. M. Jiménez de la Espada with introduction by M. del Carmen González Muñoz (Madrid, 1971). Velasco completed the text in 1574. His information was taken from various sources: numerous maps and papers left by Santa Cruz; an observation of a lunar eclipse in New Spain in 1544 by Joanoto Durán; and other astronomical observations 'made by a Spanish resident in the Philippines who is learned in mathematics,' p. 5 and p. 289.
19. *Codoin*, 2i, pp. 94–129, reproduces the correspondence between the king, Luis de Velasco and Urdaneta in 1559–61. The suggestion that Urdaneta was deceived

by the issue of secret instructions by the *audiencia* – it had assumed authority because of the death of the viceroy – is made by J. de Arteche, *Urdaneta. El dominador de los espacios del océano pacífico* (Madrid, 1943), p. 155.

20. MN:FN XVII/22/238–44v records Urdaneta's response, 8 October 1566, Madrid; *ibid.*, XVII/23/246–9v the responses of Medina and Gutiérrez, same date; BM Add. MS. 17, 625 ff. 96–100 that of Alonso de Santa Cruz, same date; MN:FN XVII/26/252–6 that of Jerónimo de Chaves, 10 October. The subsequent retractions by Santa Cruz and Medina are documented in *ibid.*, XVII/27/257–7v, 16–17 July 1567, Madrid.
21. IVDJ 25B, subsection 22, G.-B. Gesio to the king, 23 February 1576, Madrid. Against this view of spice profits, the verdict of a modern expert is that the Portuguese crown 'did not ultimately derive much profit from the cloves and nutmegs' because of the great cost of fitting out ships sent to fetch the spices and of maintaining forts on the isles; C. Boxer, *The Portuguese Seaborne Empire 1415–1825* (Harmondsworth, 1973), pp. 61–2.
22. Juan de Borja to the king, 26 November 1572, Lisbon; reproduced in G. Andrés, 'Juan Bautista Gesio, cosmógrafo de Felipe II y portador de documentos geográficos desde Lisboa para la biblioteca de El Escorial en 1573,' *Publicaciones de la Real Sociedad Geográfica*, serie B, número 478 (Madrid, 1967), pp. 5–6. This article also contains the text of Velasco's report.
23. Gesio to the king, 10 June 1579, Madrid; published in *Relaciones geográficas de Indias: Perú*, ed. M. Jiménez de la Espada, vol. 2 (new edition, Madrid, 1965), pp. 139–140.
24. Gesio to the king, 18 January 1578, Madrid; published in C. Fernández Duro, *Disquisiciones náuticas,* vol. 4 (Madrid, 1879), pp. 309–10.
25. MN:FN XVIII/21/103v–104v, Gesio to the king, 14 October 1579, Madrid. Gesio held untypical and extravagant views on Castile's territorial rights. He argued that there were grounds for believing that the whole of Brazil belonged to Castile – he referred, without going into detail, to astronomical observations 'made by Vespucci with astrolabe and quadrant;' AGI:P 29/32, Gesio to the king, 24 November 1579. And he took the eccentric view that the Philippines, though west of the Moluccas, did not come under the terms of the emperor's sale because all that had been specified was a line *east* of the Moluccas; MN:FN XVIII/80/408v, same to the same, (n.d.).
26. IVDJ 25B, subsection 22, Gesio to the king, 23 February 1576, Madrid.
27. *Ibid.*, same to the same, 24 February 1576, Madrid.
28. *Ibid.*, same to the same, 6 January 1576 and 23 April 1577, Madrid. Gesio said the conversion of Japan, like medical treatment, should proceed gradually: in the remedy of disease one quality of the body was gradually replaced by its contrary; in the replacement of the religion and customs of a land by others contrary to them drastic measures were equally to be avoided. Gamboa's interest is recorded in his brief report which is filed with Gesio's letters.
29. MN:FN XVII/80/408–10, Gesio's discourse on Macao, 1579. The king had asked him to discuss the Macao affair with Gamboa. The ambassador's confession of ignorance is recorded in *ibid.*, XVIII/15/63–63v, Juan de Zúñiga to the king, 27 December 1578, Rome.
30. Ambrosio de Morales, *Las antigüedades de las ciudades de España* (Alcalá de Henares, 1575), pp. 4v–5. Morales was the king's chronicler.
31. Felipe de Guevara, *Comentarios de la pintura* (Madrid, 1788), p. 220.
32. J. A. Maravall, *Estado moderno y mentalidad social*, vol.1 (Madrid, 1972), p. 202.
33. AGI:P 261/9, 'Información recibida en la audiencia real de la Nueva España a pedimiento de Francisco Domínguez cosmografo del rey,' February–May 1594.
34. According to one witness Jaime Juan, another cosmographer, assisted in this: *ibid.*, f. 63.
35. AGS:GA 155/149, *consulta, Consejo de las Indias,* 5 February 1583, Madrid, recommending that Juan meet Domínguez in New Spain and recover in the Philippines the papers of Fr. Martín de Rada with their astronomical observations; *ibid.*,

155/151 records the king's approval for both these suggestions, 24 February 1583, along with his instructions for the arrangements 'to be made quickly' so that Juan could sail with the next *flota*; *ibid.*, 155/150 indicates the detailed instructions issued to Juan, described as 'an expert in mathematics and astronomical calculations,' and lists with bare descriptions six instruments supplied to him to determine 'the meridian with precision,' and to measure the elevation of the pole star. He was given a free passage, a salary of 400 ducats a year and expenses of 300 ducats. In contrast to the way Domínguez was treated, there are no records of complaint about unpaid salary.

36. The new regulations concerning the cosmographer-chronicler are given in *Colección de documentos inéditos relativos al descubrimiento, conquista y organización de las antiguas posesiones españoles de América y Oceanía*, ed. J. F. Pacheco *et al.*, vol. 16 (Madrid, 1871), pp. 457–9.
37. Facsimilies of the questionnaire and instructions, and the royal *cédula* of 25 May 1576 are brought together in *Relaciones geográficas de España y de Indias impresas y publicadas en el siglo XVI*, a pamphlet with a few notes by Carlos Sanz (Madrid, 1962).
38. These instructions were sent out more than once in the 1580s; a facsimile of those sent in 1582 is reproduced in Sanz, *op.cit.*
39. Some of the *relaciones* for New Spain were published in F. del Paso y Troncoso, *Papeles de Nueva España*, 7 vols. (Madrid and Paris, 1905–6). Those for Peru can be studied in the heterogeneous collection of documents – not all generated by the questionnaire – in *Relaciones geográficas de Indias: Perú*, ed. M. Jiménez de la Espada, 3 vols., (new edition, Madrid, 1965).
40. AGI:P 260i/3, 'Relación del orden que se tubo en la ciudad de Panama en cumplimiento de lo que su magestad embió a mandar se hiziese sobre la observancia del eclipsi de la luna por una instrucción ymbiada a mí licenciado Cepeda, presidente de esta rreal audiencia de tierra firme,' 1581. The document, signed by Cepeda and Alonso Palomares de Vargas, does not make clear the respective roles of the two men in observing the eclipse.
41. Pedro de Medina, *Arte de navegar en que se contienen todas las reglas, declaraciones, secretos, y avisos, que a la buena navegación son necessarios y se debe hacer* (Valladolid, 1545), dedicatory preface to Prince Philip. Medina had been employed by the crown since 1539 as an examiner of pilots at the *Casa de la Contratación* in Seville; he continued to work there until his death (1567) and on the basis of his teaching wrote the *Arte* which was translated into several European languages.
42. Medina, *Regimiento de navegación* (Seville, 1563), p. 57. This was a simplified version of his *Arte*.
43. *Ibid.*, p. ii(verso).
44. AGI:P 259/58, Alonso Álvarez de Toledo, statement of receipt of nautical instruments, 8 January 1574, Madrid.
45. *Ibid.*, *ibid.*, 262/2, unfoliated, the king to García de Céspedes, royal cosmographer, 13 June 1596, Toledo.
46. BM Add. MS.28,339, f.237, Cristóbal de Barros to the king, 30 April 1575, Laredo, refers to the royal instruction to nominate three persons for the chair; MN:FN XXII/28/104, same to same, (n.d.) contains recommendations on the professor's salary and the duration of the course, and also reports that 'mishaps and losses' occur because of the lack of good pilots on the coast who know how to measure the altitude of celestial bodies and how to estimate the day's run. According to this source French pilots were being employed instead on low pay and 'in wartime they would not be available;' AGS:GA 210/67, *consulta, Consejo de Guerra*, 14 January 1587, Madrid indicates that local interference by the *audiencia* of Galicia was obstructing the movement of timber to Andalucía, depriving Poza of the source of his salary.
47. For details of the formation of the Academy and urban resistance to the creation of daughter institutions in the towns see D. Goodman, 'Philip II's Patronage of Science and Engineering,' *British Journal for the History of Science*, **16** (1983), 56–7.

Gines de Rocamora y Torrano, *regidor* of Murcia and himself a teacher of cosmography, looked to the Academy to remedy the shortage of pilots which caused 'the armadas, lives and honour of the world's greatest monarchy to be entrusted to foreigners;' *Sphera del universo* (Madrid, 1599), pp. 4v–5. The same source, p. 6v, indicates the names of some noblemen who attended the lectures, but the size of classes and the social origin of the students remain largely unknown. The Academy continued to function until 1625. The manuscript of Lavanha's 'Tratado del arte de navegar' has recently been transferred to the university of Salamanca library; Cedillo Diaz's manuscript 'Tratado de la carta de marear' is conserved in the Biblioteca Nacional, Madrid. Lavanha continued to serve Philip III, producing a comprehensive map of Aragón.

48. The royal legislation concerning the *Casa* is best followed in *Cedulario indiano*, the most authentic collection of documents on the government of the Indies; assembled by Diego de Encinas an official of the council of the Indies, it was first published in Madrid, 1596; a facsimile edition was published in Madrid, 1946. Philip's various enactments as regent and king relating to the *Casa* are contained in vol. 1, p. 457f and vol. 4, p. 181f.
49. He produced maps of Andalucía and of the Gulf of Mexico for Abraham Ortelius, Philip's cosmographer in the Netherlands and the author of a famous atlas; H. F. Cline, 'The Ortelius Maps of New Spain, 1579, and related contemporary materials 1560–1610', *Imago Mundi*, **16** (1962), 104–5.
50. The angular distance of the sun N or S of the celestial equator at a particular time was subtracted from the observed angle of the sun's elevation; this gave the complement of the observer's latitude. Tables of the sun's motion had been used for this purpose by the Portuguese in the fifteenth century.
51. A much simpler method which avoided the complicated motion of the sun: measurement of the elevation of the pole star above the horizon, usually by a cross-staff, gave the observer's latitude directly.
52. *Recopilación de leyes de los reinos de las Indias* (Madrid, 1681), *lib.* 8, *tít.* 23, *ley* 26.
53. *Ibid.*, *lib.* 8, *tít.* 23, *ley* 21. The existing regulations also required candidates for the pilot's examination to be 'a native of Castile, Aragón or Navarre; 24 years of age; of good morals and sound mind; neither a blasphemer nor swearer; and to have experience of six years' navigation to our Indies;' legislation of Charles V, re-enacted by Philip II, 11 November 1566, *ibid.*, *lib.* 8, *tít.* 23, *ley* 15. Apart from the crown's concern that pilots should be properly trained in the theory of navigation, there was occasional criticism (but without consequent legislation) of the lack of navigational experience of the pilot-major and cosmographers. When Hernán Pérez de la Fuente, a member of the Council of the Indies was appointed to visit the *Casa* in September 1549 he was surprised to find that the scientific staff were academics with little or no experience of navigation; letter to the emperor reproduced in *Relaciones geográficas de Indias*, ed. Jiménez de la Espada, vol. 2, p. 137. And during Philip's reign opponents of Rodrigo Zamorano, pilot-major, tried unsuccessfully to secure his removal by playing on his total inexperience of sailing. His reply was that he knew about navigation not like 'an ignorant sailor' but through a command of 'mathematics, astrology and cosmography;' and no mere sailor could perform the duties required of him as pilot-major; MN:FN XXVII/13/48–50, Rodrigo Zamorano to the king, (n.d.). It is interesting that the same point had arisen before when Zamorano was being considered for the professorship of navigation on the north coast; on that occasion pilots and mariners who were consulted thought that Zamorano's inexperience of the sea was 'no disadvantage because time alone is the master;' BM. Add. MS. 28, 339, f. 237, Cristóbal de Barros to the king, 30 April 1575, Laredo.
54. From the article on him in the *Dictionary of National Biography*. Borough was instrumental in promoting the translation into English of Martín Cortés' treatise: *The Arte of Navigation*, trans. Richard Eden (London, 1561).
55. *Cedulario indiano*, vol. 4, pp. 184–6, the king to the officials of the *Casa*, 21 October 1564 and 25 February 1565.

56. For an illustration of the seal of the *Casa* – a rare example of a surviving stamped instrument of the reign – see M. Destombes, 'Un astrolabe nautique de la Casa de Contratación (Seville, 1563),' *Revue d'histoire des sciences*, **22** (1969), 33–64.
57. *Recopilación de leyes de las Indias, lib.* 8, *tít.* 23, *ley* 30. This can be compared with the earlier allegation that ships sailing from Panama and Nicaragua had been lost because the pilots did not know how to make accurate measurements of the elevation of the pole star and sun, or how to make the necessary calculations; consequently their ships were long delayed in reaching land and lives were lost due to shortages of drinking water; AGI:P 259/48, statement by Sebastián Rodríguez Delgado, master and pilot of the South Sea, calling for the better training of pilots, 1573.
58. AGI:P 262/2, unfoliated, the king to García de Céspedes, 13 June 1596, Toledo. Together with this royal *cédula*, the same source contains various documents which record the judgments of pilots, cosmographers and mathematicians requested by the crown to report on García de Cespedes' performance in correcting charts at the *Casa* and on the quality of the standard instruments he had constructed; there was general approval and in May 1599 Philip III ordered their adoption. In one of these reports, the results of meetings in October–November 1598 at which Ginés Rocamora y Torrano, Luis Jorge de Barbuda and others considered the new instruments, the interesting comment was made that navigation to the Indies 'has not yet been given the necessary perfection so that it can be done with safety, which is desirable if other nations are not to mock Spain;' the *reputación* or prestige of Spain was at stake. The documents also record the views of Antonio Moreno, a mathematics teacher in Seville; Diego Perez de Mesa, professor of mathematics at Alcalá and Seville; and Jerónimo Martínez del Pradillo, cosmographer of Seville, on the introduction of 'curved lines' on charts to replace the usual plane charts with their straight parallel lines. It is not clear who had asked them to consider this; García de Céspedes had prepared conventional charts. Their comments confirm the usual conclusion that conservatism everywhere delayed the introduction of Mercator's spherical projection. For the part of García de Céspedes' assignment which concerned the correction of maps and improvement of instruments Pedro de Onderiz, professor of cosmography at the Academy of Mathematics in Madrid, had earlier been commissioned (1593); but he had died before a start had been made. The dispatch of external cosmographers to the *Casa* probably reflects the shortage of staff there – the accumulation of office by Zamorano – rather than any lack of skill. Zamorano was a competent and productive cosmographer: his *Compendio de la arte de navegar* (Seville, 1582) achieved five editions by the end of the century and, translated into English, it was published as an appendix to an edition of Edward Wright's *Certaine Errors in Navigation*. Around 1582 Zamorano was complaining that there was no one to help him with the demands of the *Casa* except a pilot-major who 'is almost 100 years old' (Alonso de Chaves, then about 90); and that he had worked under considerable pressure to supply Sarmiento de Gamboa's expedition to the Straits of Magellan, constructing all the instruments and navigational tables himself and going in person to San Lucar to demonstrate their use to the pilots and masters; AGI:P 262/1, unfoliated, statement by Zamorano (n.d.).
59. MN:FN XXVII/13/48–50, Rodrigo Zamorano to the king, (n.d.).
60. AGS:GA 178/384, Gian Andrea Doria to the king, 25 October 1585, Genoa. In this letter Doria told the king that despite their low qualities Spanish pilots were customarily taken on board whenever the galleys of Italy sailed along the coasts of Spain because 'they have more experience of their coasts than the best from here'.
61. M. Cortés, *Breve compendio de la sphera* (Seville, 1556), p. vii (recto).
62. They are conserved along with other documents in AGI:P 262/2; the pilots' replies are dated April–November 1597; the printed circular carries the heading: *Memoria de lo que an de advertir los Pilotos de la carrera de las Indias, acerca de la reformación del padrón de las Cartas de Marear, y los demas instrumentos de que usan, para saber las alturas y derrotas de sus Viajes.*

63. Serious errors in their use, incapable of correction anywhere in the sixteenth century, were the failure to take account of the dip of the horizon and the effects of atmospheric refraction.
64. This can be compared with Lavanha's sceptical comment: 'when navigators fail to find a place they attribute the cause of error to the current of the waters or to the variation of the magnet; but the error is due to the incorrect [representation on charts] of the relative positions of places; university of Salamanca MS, João Lavanha, 'Tratado del arte de navegar,' c.9.
65. AGS:GA 166/202, Francisco Duarte to Antonio de Erasso, royal secretary, 14 September 1584, Madrid.
66. From a statement by the pilot-major, Alonso de Chaves, in his unsuccessful attempt to prevent the employment at the *Casa* of Domingo de Villarroel, a Neapolitan cosmographer; AGI:P 262/1, unfoliated, 20 April 1584.
67. AGS:GA 283/17, Hernándo de Mendoza to the king, 11 April 1590, Pasajes.
68. *Ibid.*, 321/49, Alonso de Bazán to the king, 22 May 1591, El Ferrol.

# 3

## Technology for war

The continuing interest in exploration and the desire to control commerce with the Indies were not the only reasons for the crown's involvement with ships. During Philip's reign ships acquired a much greater military importance than before, and the king and his councillors became increasingly concerned to build and equip ships for the purposes of war. This was the consequence of expanding warfare against the Turks, and the French and English corsairs. A large fleet of Spanish galleys fought with the Holy League at Lepanto, and there had never been a greater naval battle than the engagement of the formidable 'Invincible' Armada with the English fleet. Ships were needed for other military expeditions on an unprecedented scale: for the amphibious expedition which resulted in the annexation of Portugal and for the decades of campaigning against the king's rebellious subjects in the Netherlands, which required the transport of soldiers and supplies by sea from Spain to Genoa. Castile's finances were strained beyond the limit to meet the soaring costs of war: 1½ million ducats a year were sent to the Netherlands in 1567–85, increasing to an average of 5 millions each year in 1586–90; and in 1588 the Invincible alone cost 10 millions. These sums far exceeded the income from Indies bullion and the deficit had to be supplied by various expedients: high taxation of Castilians, seizure of private money from individuals, sale of offices, and loans. Financial crises caused Philip to suspend payment of debts on four occasions (1557, 1560, 1576, 1596).[1]

The scale of war led to a more elaborate military administration, especially in the 1580s when the Atlantic became Spain's principal theatre of operations.[2] And there was a growing demand for warships, artillery, hand-guns, fortifications and munitions. This stimulated the technologies associated with their production; but in the end Spain was always short of raw materials and was forced to import at a high price from other parts of Europe. It has been suggested that this was because Philip II had no power to command or requisition in lands like Liège where the guns were made, with the result that 'the greatest ruler of the age' had to bow to 'the sovereignty of the market.'[3] But there is more to this, because at the time Spain's natural resources seemed to offer the prospect of self-sufficiency in ships, weapons and munitions. In the Sierra Morena there were numerous pyrites mines, a potentially rich source of copper, the principal raw material for bronze cannon. There were also within the Peninsula plentiful supplies of saltpetre earth and some sulphur, the chief ingredients of gunpowder. There were extensive forests providing timber suitable for shipbuilding, including pine masts, which although not equal to the best from the Baltic, were still serviceable. In the moist zone around Calatayud large quantities of hemp were grown and used for ships' rigging and arquebus match. Yet Spain regularly imported bronze cannon from Flanders and Germany; copper from Hungary; gunpowder and sulphur from Italy; masts from the Baltic; and hemp from the Netherlands and Germany.

What had gone wrong? Why this failure to supply strategic raw materials and manufactures urgently needed for war? Were the king and his councils negligent in exploiting natural resources, or were they facing insurmountable technological obstacles? And, more generally, what steps were taken by the king and his councils to promote technologies for the development of Spain's military power in an age of intensifying warfare?

## *1 Philip II and shipbuilding*

Historians used to say – it is difficult to say why – that Philip II failed to appreciate the importance of naval power, displaying 'an indifference, or even apathy, toward everything connected with the sea, which was one of the unpardonable mistakes of his policy.'[4] Philip's legislation on the conservation of forests is sufficient to show that this conclusion is unwarranted. Spanish monarchs since the end of the fifteenth century had been concerned with the depletion of forests, but chiefly for reasons unconnected with shipbuilding. There had been royal edicts restricting cuts to the branches of trees in extensive forests (1496), and ordering local authorities to inspect forests and restore them by making new plantations (1518).[5]

The intentions here were to maintain acorn feed and shelter for cattle, and supplies of firewood. Not till 1547, when Philip was regent for the absent Charles V, was there royal legislation on forests specifically directed to the interests of shipbuilding.[6] The preamble stated that there had been excessive cutting of trees in Guipúzcoa and Vizcaya, provinces where 'many ships are fabricated for our service', and that as a result there was a shortage of timber for shipbuilding. To remedy this, it was ordered that no person in these regions 'may cut a tree without planting two others' and all who had felled trees in the previous ten years were required to replace the losses with fresh plantations.

But it was in the 1560s and 70s that Philip II revealed the strength of his determination to maintain timber for shipbuilding. To his representatives, the *corregidores*, in Guipúzcoa, Vizcaya and the Cuatro Villas de la Costa de la Mar (the maritime centres of San Vicente de la Barquera, Santander, Laredo and Castro Urdiales) he issued instructions to implement an elaborate plan to rectify the 'great shortage' of ships caused by continuing 'negligence in conservation and plantation of forests.'[7] Although the king's letter spoke only of the damage to commerce and the impoverishment of local populations, he was also concerned with naval needs. There was no clear distinction at the time in Spain (or England) between merchant ships and warships; merchant ships were commonly requisitioned for military use. The *corregidores* were ordered to inspect sites for plantations 'two leagues from the sea or from navigable rivers' and assess the number of oaks to be planted there each year. This estimate was to be communicated to municipal authorities (*alcaldes* or *justicias ordinarios* and *regidores*) who were then required to apportion a definite number of oaks for planting by individual landowners or, in the case of communal land, by the local corporation. The precise figures allocated to each district were to be recorded in a register for the *corregidores* to consult each year, when they received the mandatory annual reports of actual planting from local justices. Failure to fulfil the numbers was punishable by fine: one *real* from landowners for each tree that had not been planted, and the same from each offending municipality, the money being taken from the personal funds of *alcaldes*. Measures were also to be taken to ensure that the new plantations and existing forests would be protected from cattle.

The king meant business and asked *corregidores*, within thirty days of receiving these orders, to send a detailed report on what steps had been taken. But the enforcement of these regulations proved to be no easier than in contemporary France, where royal forest laws went unheeded by powerful feudal proprietors and peasants who continued to destroy forests by grazing and pasturing their flocks. Philip also encountered local opposition and was soon informed that his instructions had fallen on deaf ears. *Corregidores* and justices had done nothing and were now threatened with heavy fines.[8] Part of the inaction was due to the king's failure to provide money for the scheme; and he now suggested that this might be raised by local taxes on provisions, a means which could also furnish the salaries of forest guards. Nor had the justices been given pay for their additional task; they 'preferred to make money in the towns than spend it visiting forests',[9] and eventually the king was forced to offer them payment. There was also the problem of land ownership – much of the land in Spain did not belong to the king, and when plantations were ordered along the coast of Galicia confrontation was envisaged with numerous seigneurial and abbatial proprietors.[10] In an effort to overcome all such local opposition Philip appointed a royal official with extraordinary powers to enforce the forest laws. Cristóbal de Barros became *superintendente* of forests and plantations.[11] His activity prior to this is unknown and his appointment simply referred to him as a servant of the king with suitable experience. He was now empowered to inspect the accounts of corporations who pleaded poverty as the reason for their failure to plant oaks, and if verified Barros could impose a local duty to raise the necessary money; he could enforce the appointment of forest guards; review cases of transgression of the laws protecting forests and enforce penalties when local justices were lax; dismiss uncooperative justices, order arrests and punish 'as a justice does'. He was even given the authority to supervise *corregidores*, reporting any negligence to the king.

Barros was to be answerable to the king alone, and *corregidores* and justices were told not to interfere. When subsequently judicial authorities in Galicia did try to interfere after Barros had levied an excise at Vivero, and again when he had made arrests at La Coruña for failure to pay a tax he had imposed for plantations there, the king sent a stern rebuke ordering them to cease obstructing the plantations.[12]

When Barros and his assistants at last began planting, the

techniques used do not seem to have been successful. The long-term policy of planting seeds or saplings appears to have been avoided, probably because the king wanted timber quickly. Instead growth was encouraged by thinning out existing forests, transplanting mature trees to new ground. According to one local complaint this had several unfortunate results. The transplanted trees had shrivelled and died; they had been planted on communal land used for wheat cultivation; and populations, with less timber and bread than before, were refusing to continue transplanting.[13] An official investigation later confirmed that this was why the plantation scheme had failed in the Cuatro Villas de la Mar, and that some of the residents had fled, unable to pay the fines imposed on them for refusing to plant.[14] Although Barros wanted the plantations to continue, the Council of War recommended their abandonment and advised instead that conservation alone would be sufficient to meet the needs of shipbuilding; there is no record of the king's decision. But the conservation plans had also run into difficulties. In 1574 the king had prohibited the felling of oaks 'unless it may be for building a ship or a house' along the entire northern coast from the French frontier to the border with Portugal. Reports from Barros and others showed that this was not practicable and the king had to relax the prohibition and permit the cutting of oak trunks for the construction of bridges, ploughs and carts, and for the supply of timber for vineyards.[15] However great the need for ships these other necessities of life could not be ignored.

Royal authority also intervened in the protection of the pine forests of Catalonia, the other principal centre of shipbuilding in the Peninsula. There forests were being consumed to supply fuel for furnaces for glass manufacture. The ruling Council of Twenty in Barcelona had called for the dismantling of the furnaces which threatened to halt the construction of galleys for the king; but vested interests seem to have prevented this.[16] The endemic problems of Catalan brigandage also had its effects on the forests – trees had been cut to deprive bandits of hideouts.[17] An initiative originating from Hernando de Toledo, captain-general of the Principality, brought some progress. His proposal that Juan Comalonga be appointed to supervise plantations by private landowners was recommended by the Council of War and approved by the king. It was reported that 'many plantations' were accomplished in 1578–

9; but after that nothing had been done for ten years, because Comalonga had become occupied in legal affairs, and – more to the point – he had received no pay from the king. The Council of War repeatedly pressed Philip to pay what was due; but plantation work was not resumed until the 1590s under the Maestre de Montesa.[18]

The supply of timber for galley construction at Barcelona had also become more difficult because of grants of privileges, described as 'recent' and 'numerous', to monasteries and individuals. As a result the crown no longer had cutting rights in the forests of Sabadell and other localities within a few leagues of Barcelona, which had traditionally supplied the royal arsenal. Distant forests had therefore to be exploited instead, notably those around Montseny and Arbúcias, four leagues from the coast at Blanès and another nine leagues from there to Barcelona, bringing great increases in transport costs.[19] That goes some way towards explaining Spain's shortage of masts, at least for galleys. In May 1586 2860 trees, including 886 pines, lay on the ground in the forests of Montseny, deteriorating because the 1430 ducats needed for their transport by land and sea to Barcelona had not been supplied by the king.[20] This was a recurring source of irritation at the royal arsenal. When in 1589 2400 sawn logs of pine, destined for various large pieces of galleys, were abandoned in the same forests, Gerónimo Urpín, a carpenter, was contracted to transport the timber to the coast; but he used the money to pay off his debts and two years later emergency action had to be taken to retrieve remaining pines which had not yet rotted and bring them to the shelter of Barcelona's arsenal.[21] And later when the source of pine masts at Arbúcias seemed to have been exhausted, searches further afield reached out to the extensive pine forests of the Pyrenees. In one forest, on land belonging to the king in the viscounty of Castilbo, a well-informed official said there were 'so many masts and spars that it would not be exhausted even if many galleys were constructed each year.'[22] But this was on the borders of Andorra, in mountainous terrain and entailing still greater transport costs. The viceroy of Catalonia ordered a road to be constructed from the forest to the river Segre, two leagues away and for the most part over flat ground. The pines were transported along the river (the sources are not clear if this was done by river craft or, more likely, simply by floating the

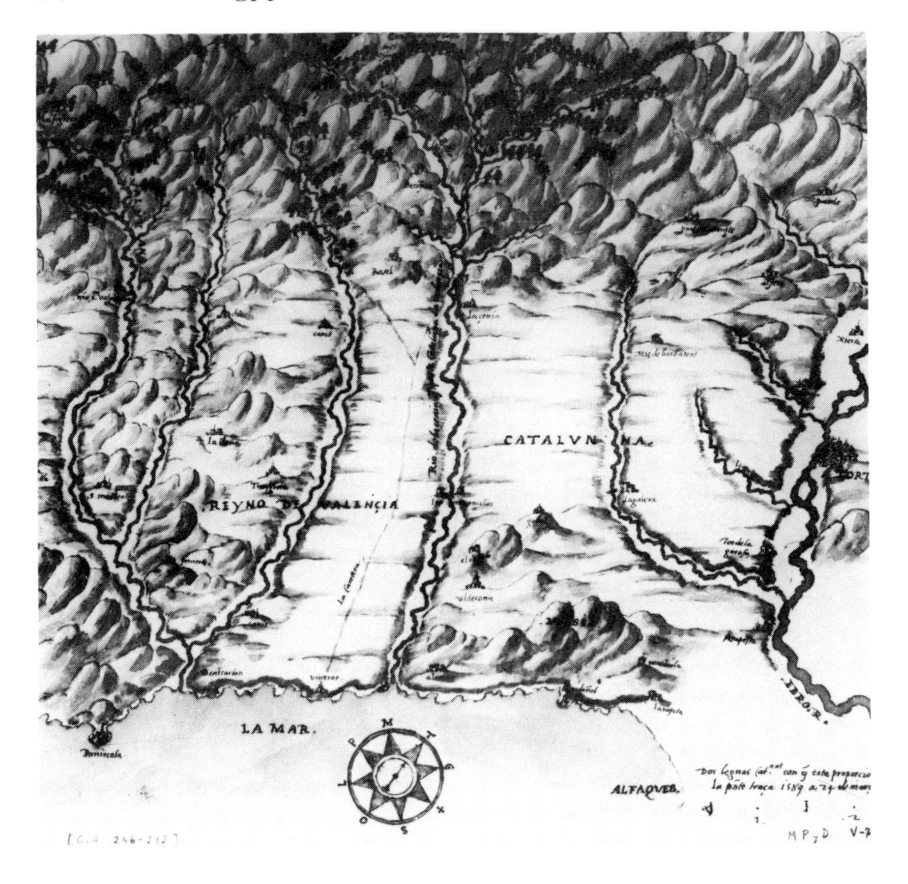


Plate 2 Map of a forest region on the borders of Catalonia and Valencia sketched by Giorgio Setara, an Italian engineer, who accompanied an official and master craftsman from the royal arsenal at Barcelona in a reconnaissance of an area reported to have rich timber resources for shipbuilding. In the middle can be seen Setara's suggested route for a new road to transport timber to the coast at Vinaroz.
(AGS:GA 246/282, *Mapas Planos y Dibujos* V–79; 24 March 1589).

logs) which joined the Ebro and so reached the coast at Tortosa where they were shipped to Barcelona. This route had its difficulties. The middle section of the Ebro had a rock and at that point the timber had to be put on land and then laboriously returned to the river. There was also danger from Barbary corsairs lurking off Tortosa. The costs were high, and soon after the first cuts had been made there were complaints from Barcelona that while a few of the 30 masts and 26 spars had arrived, most lay in the Pyrenees because of lack of money.

The importance of transport in forest exploitation was fully appreciated by the king and his ministers. When news reached the king that there were beech trees in Galicia suitable for making long oars for galleys he wrote to the governor for detailed information on the distances of the forests from the coast, envisaging carting and then shipment to Andalucía.[23] And when optimistic reports came in of rich timber resources on the borders of Catalonia and Valencia, the viceroy of Catalonia, Manrique de Lara, ordered by the king to arrange an inspection, sent to the area a small survey team whose composition indicated that the practicalities of transport were uppermost in his mind. Along with Antonio de Alzatte, a Neapolitan then supervising galley construction at Barcelona, and Juan de Nicola, a Ragusan master-builder of galleys at the same arsenal, who were sent to examine the size and flexibility of the trees, went Giorgio Setara, an Italian engineer, 'to see if there is a road or if not how one could be constructed and at what cost.'[24] Setara prepared a map and reported that two roads would be required: one six leagues long, easy to construct because the terrain was flat; another, more difficult and two leagues long, would cost 1500 ducats. But the overall report was unfavourable, finding that the pines were 'weak and porous and more appropriate for houses or small boats than for galleys' and the oaks inadequate for wales, futtocks and long pieces needed for galleys (plate 2)[25].

Masts from the Pyrenean pine forests reached Barcelona with difficulty. There were no illusions that the Pyrenees might also supply masts for the galleons being constructed on the Atlantic coast. But after the annexation of Portugal (1580) the rich pine forests of the Algarve became available and might conceivably have been used for that purpose; in 1582 21 masts were bought at Faro and taken to Cádiz.[26] Was it the mountainous terrain of the Algarve which prevented Spain's systematic exploitation of these forests?

Whatever the reason the annexation made no difference to admirals and shipbuilders; they took it for granted that masts for galleons would have to be imported from the Baltic.[27]

During Philip's reign timber had acquired strategic importance. The king introduced severe penalties for exporting prepared timber pieces suitable for the parts of ships.[28] The ambassador in London reported with dismay that the English were supplying large quantities of worked timber for galleys to the king's enemies in Barbary.[29] And when Luis de Requesens, lieutenant-general of the galleys of Spain, sent his congratulations to the king for the victory at Lepanto, he advised the burning of forests used by the Turks to prevent the restoration of their fleet.

Barcelona was the centre of galley construction in the Peninsula. The late-medieval *atarazana* (arsenal) with its arched bays belonged to the monarch. It had declined with the loss of Aragón's Mediterranean trade in the fifteenth century, but was temporarily reactivated by Charles V's order for 50 galleys for the expedition to Tunis (1535). When he became regent in 1551 Philip provided a bastion to defend the exposed arsenal from French attack. And during his reign he intended to keep the *atarazana* in a state of continuous activity, furnishing galleys, in addition to those produced in the other royal arsenals in Naples and Sicily, for the defence of the western Mediterranean against the Turks and their allies, the Barbary corsairs. The need became acute in 1560–2 after two naval disasters had halved Philip's galley fleet: an abortive attempt to recapture Tripoli ended at Djerba where the surprise arrival of the Turkish fleet resulted in the loss of 28 galleys; and a further 25 galleys were lost and thousands of men drowned during a storm in the bay of Herradura on the coast of Andalucía. Yet within a decade and at a cost of some three million ducats the king was again in a position to meet the Turk – the result of efficient organisation of craftsmen and materials at Barcelona by the viceroys of Catalonia.[30]

The documentary sources of information about construction at the *atarazana* are fragmentary for the early part of Philip's reign; but the record is fuller for the 1580s and reveals the host of difficulties which hindered the fulfilment of orders for the completion of a specified number of galleys each year. In fact it was because all was not well at the arsenal that more detailed information exists – the reports of official investigations. By this time Spain's attention had turned to war in the Atlantic and the Turks were preoccupied in the east with Persia. But the threat of the Turkish fleet continued to be felt, and to the end of his reign Philip was anxious to maintain his fleet in a state of readiness at Barcelona and to continue the production of new galleys.[31]

During the winter of 1586 severe storms had prevented the shipment of timber from Blanes and Tortosa to the arsenal, and plans to boost production by employing additional craftsmen had to be dropped because there was not enough timber for them to work.[32] The king's indebtedness was a more frequent cause of interruption to galley building; failure to provide money for transporting logs from the forests or for the workforce at the arsenal soon brought construction to a halt. At the end of 1587 Manrique de Lara informed the king that 'all had become still' in the arsenal and 153 carpenters and caulkers had been dismissed, leaving only 30. He had written several letters to the king asking for money, but had received no reply. There was nothing with which to buy oakum or pitch, or to pay men cutting and sawing in the forests, transporting logs, manufacturing iron nails, and constructing galleys. The lost time could not be made up; the viceroy said that even if the king should send a large sum of money immediately, it would be too late to permit a single galley to be launched in the following summer. And during that summer, because the arsenal was again inactive, the viceroy gave one of the master craftsmen two months leave in Madrid, sending him on a circuitous route by galley to Cartagena so that from his own experience of sailing in them 'he could better understand what needs to be done to modify or perfect' the galleys he was building.[33] The only way work could begin on constructing eight galleys for completion in 1589 was for the viceroy to take extraordinary action; he diverted 15 000 *libras* which had been voted by the *Cortes* of Monzón for repairs to the fortress of Perpiñán and other fortifications, and used the money to cut, saw, and transport timber for galleys, a decision which brought continuing protests from the *Cortes* up to a year later.[34]

Even when money from the king did arrive it did not go very far. The viceroy was 'astonished' at the costs of galley construction and, suspecting 'theft or negligence,' wanted an investigation. He told the king that the six galleys due to be built in 1589 ('if money shall arrive') would each cost 4000–5000 ducats and 'I have always thought this scandalous.' And when Antonio de Alzatte was brought over from Naples to improve the running of the royal establishment he too was shocked by the high costs at Barcelona.[35] Some of the excessive costs were due to inefficient organisation. Juanot Sangles, commissary at the arsenal, was reprimanded for appointing 'an inept person who could neither read, write, nor keep accounts of the quantities of carts and timber' which arrived.[36] There had for a long time been inadequacies in recording the entry of raw materials within the arsenal and their subsequent distribution. Stocks were supposed to be protected by a guard of fifty soldiers; but their captain was found to be organising gambling within the building, there had been no watch kept at night and timber, nails and other supplies valued at over 1000 ducats had been stolen by the soldiers themselves.[37] And Bartolomeo Giordano, one of the master craftsmen, was one of several arrested on a charge of defrauding the king's treasury; he confessed to stealing materials from the *atarazana*.[38]

Tensions existed between the administrators of the arsenal. When Pedro de Isunza, *veedor* (senior administrator) of the arsenal ordered wooden blocks and props used for launchings to be moved into the shelter of the interior, there was a squabble with the viceroy over who had the authority to give this order.[39] Worse, two of the master craftsmen had come to blows over precedence. The trouble began in January 1585 when Pedro Catalan arrived from Seville (or Portugal – the sources are contradictory), retaining his former salary of 450 *reales* a month. This was double the salary of Bartolomeo Giordano, yet he had the title of *maestro mayor* of galley construction. Catalan and his Ragusan companion, Juan de Nicola who had come with him, were each described in lesser terms as 'masters of making galleys.' Isunza saw 'three heads in one house' and sensed trouble. He arranged a meeting with the three men and got them to agree that since Catalan was more highly paid it was the king's wish that he should have greater authority and that 'the other two should obey him like a father, and he should treat them like his sons.' But this didn't settle matters and Isunza soon reported that Catalan was causing trouble by his arrogance and accusations – later borne out – that Giordano had committed theft; Giordano in turn accused Catalan of hiring a bandit from Tortosa and three other villains to kill him. Each accused the other of firing shots. The likely consequences of this enmity between two of the principal craftsmen at the arsenal were indicated by Juan de Carlua, who had been sent to investigate complaints reaching the Council of War concerning the quality of galleys constructed at Barcelona. Carlua reported that Giordano did not have his mind on the job and 'does not always use correct proportions when he measures out galleys.' He had witnessed rivalry between craftsmen on other occasions and was sure 'it has been damaging to Your Majesty's service.' Although he recommended the separation of the two men this was not done until three years later when Giordano was dismissed for theft and Catalan assumed the title *maestro mayor.*[40]

Alzatte found various other abuses at the arsenal and in his report to Andrés de Alva, secretary of war (sea), he recommended reforms for greater economy and efficiency.[41] The use of weak timber, cut out of season and green should be stopped; this had resulted in inferior galleys. The indispensable iron nails – around seventy *quintales* (hundredweights) were needed for each galley – were often not available at the arsenal because they were manufactured at numerous places along the coast up to Palamos; instead manufacture should be concentrated within the arsenal to maintain stocks. Royal officials had lost credit with the local population by not keeping their promises of payment for work done; they should be replaced by private individuals contracted to coordinate cutting, sawing and transport. Justices should be alerted to rectify current abuses in which men engaged in the transport of timber were purposely moving slowly because they were being paid by the day. Materials entering the arsenal often disappeared; detailed records and checks were essential if the cost of galley construction was to be reduced. Master craftsmen were absent without leave and an absurdly inadequate method was used to record the hours worked: a foreman took a rollcall of the workers on Saturdays so that presence on that day alone would be sufficient to guarantee a week's pay. There were too many assistants and unskilled craftsmen on full pay. Alzatte thought that the permanent labour force at the arsenal need be no more than 30 craftsmen, working continuously to produce beams, wales, futtocks, and other components of galleys, so that when a naval emergency arose, several galleys could be assembled quickly. To increase efficiency he also recommended extending the covered area so that craftsmen could continue working when it was raining.

There is no record of the formal implementation of these

reforms. Pilfering seems to have continued. A year after Alzatte's report the *veguer* (chief magistrate) of Barcelona informed the king that 'the *atarazana* is very badly governed; news reaches me every day of thefts of gunpowder, lead, copper utensils, grease, tar and other things' and also that during the recent pestilence, which had brought the arsenal to a halt, the guard of soldiers had burned a quantity of timber said to be enough to build three galleys.[42] But some improvement resulted from *asientos* (contracts) with private individuals along the lines suggested by Alzatte; they seem to have facilitated timber supplies to the arsenal in the early 1590s. And in these same years there may even be evidence that the costs of galley construction were falling. Alzatte estimated that to construct a galley at Barcelona in 1589 would cost 41 047 *reales*, not including anchors, barrels or tackle; yet in the period August 1591–August 1593 when twelve galleys were built at the arsenal they cost a good deal less – 37 532 *reales* each including tackle.[43] But the decreasing output – just twelve galleys in two years – compared with the fifteen built in 1587 reflects the decline of the *atarazana* which was accentuated in the early seventeenth century.

The high costs of labour at the arsenal, a constant source of complaint, was also indicative of the shortage of local skills; much of the excessive cost was caused by the regular import of craftsmen from Genoa. In 1565 García de Toledo, captain-general of the sea and a former viceroy of Catalonia, was convinced that there were no masters in the Principality with the technical skills needed to produce galleys according to specification; hence his annoyance when he discovered that incompetent Catalans were being employed, out of sentiment, as master craftsmen in place of the Genoese – 'the best that there were' – whom he had sent at considerable expense.[44] In 1575–95, of the six regularly employed master craftsmen, only one, Pedro Catalan, was a Spaniard; the others tended to be Genoese, like Giordano (and his father before him). During 1587 as many as 153 Genoese carpenters and caulkers had been brought to the arsenal.[45] And when a new Royal Galley was planned at the end of 1597 a special team of technicians was imported from Genoa: two masters, Geronimo Verde and his son Bartolomeo; six carpenters; a caulker and four assistants.[46] Some indication of the value placed on Genoese technical skill can be seen in the aftermath of Bartolomeo Giordano's conviction for fraud. The viceroy had ordered his dismissal from royal service; but Isunza, the *veedor*, was trying to persuade him to remain in Barcelona because he and a nephew leaving with him, Francisco Gandolfo, had skills in galley construction too valuable to lose. Isunza therefore appealed to the king to review the case; but on this occasion peculation was not excused and the judgment stood.[47]

To reduce the costs of building galleys the king was advised to imitate the Turk and use slaves to cut timber, saw planks, make nails and sails, and renew planking in old hulls.[48] The proposer had the Naples arsenal in mind. And there slaves in large numbers were seen to be an essential part of the establishment; in 1578 the viceroy of Naples told the king that construction was progressing slowly because galley slaves had not yet been delivered, and that 'at least 160 to 200 would be needed.'[49] In Spain thousands of Mohammedan slaves became available after the suppression of the Morisco uprising in Granada (1568–71) and the defeat of the Turk at Lepanto (1571). There is no evidence that any were brought to work at the Barcelona arsenal; they became gally slaves, worked in the mercury mines at Almadén or entered domestic service. But during the reconstruction of the galley fleet in the 1560s convicts condemned to serve on galleys were ordered to begin their sentences with a period of labour at the arsenal.[50]

Attempts were also made to subject the population to forced labour at so-called 'fair' rates of pay. The king advised persuasion rather than force when he instructed *corregidores* in Vizcaya, Guipúzcoa and the Cuatro Villas to send 200 of the best carpenters and 100 caulkers to Barcelona, recommending the use of 'well-chosen words so that they go of their own accord' and giving money to them for the journey and to the women they left behind.[51] But on other occasions Philip ordered the compulsion of carpenters, caulkers, smiths, ropemakers and labourers, requisitioning them along with the beasts and carts seized for transport.[52] In 1566 the viceroy of Catalonia ordered all carpenters and caulkers along the coast of the Principality to assemble in Barcelona, and when they resisted, because they did not want to leave their families and occupations, the viceroy ordered the use of force.[53] Reglá's discussion of this incident leaves the impression that Catalans could be forced to work at the arsenal. But this was not the case. When a later viceroy, after repeated requests for 15 000 ducats to complete fifteen galleys, revealed that the Catalan workforce had been 'deceived' with promises into working for six weeks without payment, he told the king in the clearest terms why he had been unable to keep the men any longer:

The people of Catalonia, by the liberties which they have been granted, are not like those of the kingdom of Toledo where an *alguacil* [constable] can order carpenters to be brought by force. Here Your Majesty is seen as an individual in a contract. They do not want to work any more and they have gone back to their homes or to other places to earn a living.[54]

Catalan freedom was also manifest at this time in resistance to royal orders in the forests. The men of Santa Coloma, La Esparra and Castanyet, unpaid for their cutting and sawing of timber for the royal arsenal, declared that 'they were no longer under any obligation to serve Your Majesty and would not permit the removal of timber from the district unless they were first paid their arrears'.[55] Grievances were also voiced at the *Cortes* of Catalonia that the viceroy was acting against the liberties of the Principality in trying to force peasants to transport timber under unfavourable conditions.[56] The royal commissary-general of Catalonia might boast of powers to compel men to work at the arsenal under penalties of fine or imprisonment;[57] but the reality was otherwise. The king was not master of Catalonia and could not do as he pleased. Nor did he have the power to compel craftsmen in the Basque provinces where long-established *fueros* (liberties) had to be observed by the monarch. At Deusto on the river of Bilbao where galleons were being built during 1589, the royal official in charge of construction had to use his own money to keep craftsmen at work on the site, and eventually when royal funds still failed to arrive, he had no option but to let the men go because 'I could not detain them'.[58]

The techniques of construction at Barcelona's *atarazana* were those which had long been used for the traditional galley. But one noteworthy innovation may have been introduced during Philip II's reign and if so it could well have been in advance of arsenals elsewhere in Europe. A document of 1586 describing stocks held in the *atarazana* mentions that oak timber was being kept in a *foso* (pit or ditch).[59] More information on this comes from a contemporary account of the king's visit to the arsenal in March 1585 after attending the opening session of the *Cortes* of Aragón at Monzón. Philip witnessed the launching of two galleys destined for the

Indies, inspected work in progress within the arsenal, and finally went to the *foso* to observe how oak timber 'was kept in water.' This suggests that a process of water-seasoning was being employed, a faster method of removing superfluous sap and moisture than the natural method of drying in air; water from the ditch would have entered the pores of the timber, dissolving the sap and forcing it out, after which the timber would have been left to dry. The technique, attributed to Isunza, was said to be 'very successful'.[60]

So long as the Turks employed galleys the continuing production of these ships in Barcelona made sense; round ships of sail would have been immobilised by the Mediterranean calms, and at the mercy of Turkish galleys armed with cannon at the prow and avoiding broadside fire. Similar conditions existed in the coastal waters of the Caribbean, where calms or prevailing breezes hindered sailing ships, and where warfare, as in the Mediterranean, was amphibious. And so in response to a request from the authorities in Hispaniola in 1577 for galleys to patrol the island to prevent corsair seizures of skins and sugar, and to suppress negro rebellions, two galleys were constructed in Spain and sent out to good effect. But for the turbulent waters of the Atlantic galleys were wholly inadequate, unable to cope with the towering waves and strong gales. The galleon, invented around 1520 in Venice or developed from the Portuguese carvel – it is uncertain which – was soon adopted by Spain. Alvaro de Bazán, Charles V's 'captain-general of the ocean sea', contracted by the crown to defend Spain's Atlantic coastal waters 'from the straits of Gibraltar to Fuenterrabía' from the French enemy, supplied a fleet which included two galleons of 'new invention'.[61] And to deal with the growing menace of French and English attacks on the shipping routes to the Indies twelve galleons were built at Bilbao in 1567–8 to the specifications of their commander Pedro Menéndez de Avila who had designed them. He had provided them with oars, seeking to combine the seaworthiness of galleons with the mobility in calms of galleys.

The discussion of ship design in Philip's reign was stimulated much more by the round ships than by the galleys of traditional form. The optimum size of vessels was debated. It was alleged that large ships were less safe because they had more joints and were therefore more prone to break up during a storm.[62] Optimum proportions were also discussed. Menéndez's galleons were experimental: the depth of hold was reduced; they were slimmer and longer, increasing the ratio of length to beam from the conventional 3 : 1 to 3.5 : 1; and they were without the usual castles.[63]

Considerations of capacity and proportions became prominent when the king sent Cristóbal de Barros to activate shipbuilding 'for defence and commerce' all along the Cantabrian coast. The appointment, made in February 1562, gave Barros a salary of 100 ducats a month[64] and put him in control first of financing shipbuilding, then of registering newly-launched ships and sales to prevent their export; finally, from about 1577 to 1589 he was put in charge of the entire organisation of shipbuilding in the region. All this in addition to his considerable powers as superintendent of forests and plantations. Barros went to a region with a long-established maritime activity. In the late Middle Ages the men of Vizcaya and Guipúzcoa had taken their round ships to the Levant, and later to the Newfoundland fisheries. They had also carried wool exports to northern Europe until the recent decline of that trade. Now the region's experienced craftsmen and shipbuilders were to be encouraged to increase construction of merchant sailing ships and produce galleons for the defence of the Indies route. On the advice of Barros the king exempted shipbuilders (April 1563) from the *alcabala*, the 10 per cent sales tax which oppressed all commerce in Castile, in sales (within the Peninsula) of ships of 200 *toneles* and above; and similarly removed the *alcabala* from sales of timber, nails, hemp, anchors and all other supplies for ships of this capacity.[65] The king also extended the system of royal subsidies, paying shipowners 10 000 *maravedís* per hundred *toneladas* for vessels over 300 *toneladas*.[66] In May 1563 the king authorised the release to Barros of 20 000 ducats for interest-free loans to shipbuilders, and by 1570 this had been used in 60 contracts. But the system of loans, seen by the king as a fund circulating after repayment, got into difficulties when shipbuilders failed to repay on the agreed date, and when justices took no interest in collecting debts for which they were given less than a tenth of the sum recovered.[67] Consequently the king lost patience and began imposing fines on debtors.

Barros opposed the merchants of Bilbao who had a strong preference for small ships. They had wanted ships to be built of 25 to 100 *toneles*, because loading could be completed more quickly, preventing the decay of merchandise; also the costs of insurance were lowered by distributing the merchandise in several small vessels; and smaller ships were less impeded by shallows or the bars at the entrances to ports. But Barros, looking beyond mercantile interests, insisted that ships like this would have 'no force of offence or defence.' The crown needed ships 'fit for war, trade and commerce,' and that meant much larger vessels of 400–700 *toneles* capable of carrying artillery and hundreds of men.[68] Although small ships continued to be built this was not to be assisted by royal loans.

Tonnage and relative dimensions were an important part of the discussions Barros had with seamen and craftsmen in 1581, when he was asked by the king to consult experts on the design of new galleons for the guard of the Indies.[69] Meetings were held in Santander and Seville to consider if modifications to Menéndez's galleons were necessary. That was agreed; the galleons had been ineffective in engagements with the enemy. Their slimness and especially the reduced heights of their decks had left inadequate room for provisions and munitions. Because material had been piled up over the decks, there were fatal delays in clearing them to allow the artillery to be brought into action. The new galleons would have to be more capacious. But they would also have to be swift enough to outpace the corsairs. There was some disagreement on how that was to be achieved. Pedro Sarmiento de Gamboa, an experienced naval officer, advised slimness. Barros recommended greater length because longer ships 'do not rise and fall with the waves, and the sails receive more wind when the masts are further apart.' Unlike Sarmiento, he thought increased breadth would be conducive to speed since there would be room on the hull for extra sails.

They were groping with a complicated problem on which little progress was made anywhere until the late nineteenth century when reliable information on optimum proportions was first acquired by extensive experiments with small-scale models. But the discussions in Philip II's Spain reveal an open mind and a willingness to learn from experience, and certainly not inflexible conservatism. In the end the galleons were built longer, broader and deeper than those of Menéndez. In November 1581 the king gave his approval for the construction of nine galleons: two, including the flagship, of 56 cubits length, 16 breadth and 400 *toneles*; and seven smaller vessels, 52 cubits long, 15 broad and 300 *toneles*. The tonnage had been kept down to facilitate rigging and to ensure that the galleons would be 'less submerged' and able to enter rivers.

Barros was put in charge of construction and ordered by the king to ensure that the galleons 'remain true to plan.' Barros replied that there could be no guarantee of the final tonnage because of inevitable errors in the techniques of construction: he said shipyards were usually situated on marshy land and the props supporting the rising hull sank under the increasing weight as construction progressed, so extending the vessel and enlarging the final tonnage. In the end the galleons were a good deal larger than intended, the six smaller ones averaging 510 *toneles*. Barros was instructed 'to proceed with the gratest possible speed,' to select suitable sites, secure the necessary workforce, purchase well-seasoned timber strong enough to withstand the 'torments of artillery,' requisition carts and animals; and finally to send to Seville nine galleons, fully-fitted, equipped with artillery, stocked with provisions and manned by seamen and pilots. He decided against simultaneous construction on several sites. That would create transport difficulties because of the 'very bad' roads along the coast and it would be harder to supervise the workforce, an impossible task at the best of times because of the large numbers of men involved. He was determined to minimise shoddy craftsmanship: the careless caulkers whose hasty insertion of oakum resulted in leaky ships; the careless carpenters whose poor scarfing reduced the strength of ships; the careless nailing which failed to match the sizes of holes to nails; and the careless manufacture of the nails themselves – he said that if a *vitola*, the thin rod used to measure the depth of a hole, was used to specify a required length of nail, the hundred or so ironworkers would never make them of uniform size. Construction on a single site would have the additional advantage of using surpluses of material for galleons in different states of construction. Barros chose Guarnizo, two leagues from Santander. Its waters presented no obstacles to navigation, it offered a secure mooring for launched vessels, and there was 'a mine of forests' of oak in the locality which belonged to no individual; in taking the timber Barros assumed 'Your Majesty has no obligation to pay, nor have I done so'.[70]

Everything was to be coordinated. While gangs of men were cutting timber, others were to begin making nails, preparing hemp and oakum, spinning and twisting cables; and searching for masts, munitions and artillery. Barros told the king that a successful outcome required more than his own diligence and divine assistance; he must have money to pay the workforce and purchase supplies. He was obliged to make repeated requests for money throughout the two years of construction. The project could only be initiated by Barros borrowing 2000 *escudos* against his own sureties to pay carpenters and sawyers; and wages at this time were rising with prices. In January 1582 he informed the king that 'unless money comes soon the construction of the galleons will not be possible'. But it was not until June that money first arrived, and even then he could not keep it – 'the Council of War asked for it and most of it was diverted'.[71] Little wonder that for his task 'the worst thing was the uncertainty of money'. He was experiencing the effects of the drainage of the king's resources by the war in the Netherlands. The king's soldiers and seamen were left unpaid, but eventually the importance of the Guarnizo project brought Barros most of the money he needed. By February 1583 he had received 43 500 ducats and was asking for another 30 000 to pay the workforce which now amounted to 'over 350 excluding carters', and to purchase imported masts, pitch and hemp. Some of the hemp had been supplied from Calatayud (Aragón); the quality of cables made from it was praised by shipowners. Yet Barros also imported some because of its high price in Spain. Was this due to high wages, to shortages resulting from disappointing harvests, or to the profiteering in naval supplies in the 1570s, when Barros had complained to the king that craftsmen were buying up hemp, pitch, tar and oakum for resale to the shipbuilders?[72]

The first of Barros' galleons was launched on 19 May 1583: the *San Felipe y Santiago* and *San Meterío y San Celedon*. He had asked for the ships to carry the names of saints 'because this is the most notable shipbuilding achieved by any king of Spain'.[73] The launched galleons remained at Guarnizo still requiring artillery and munitions for their completion; soon they were joined by the *San Pedro, San Juan, Santiago el Mayor* and *La Ascension*. Throughout 1583 Barros begged the king to supply the guns and a further 20 000 ducats needed to complete all construction, to pay what was owed to the ropemakers, and to feed the guard of fifty soldiers who slept aboard the new galleons. The artillery had still not arrived in November when reluctantly Barros sent the unarmed *San Christóbal, Sanctissima Trinidad* and *Nuestra Señora del Barrio* on their maiden voyage to Seville; within a few months all nine had been delivered. They eventually received their guns and served first to guard the Indies fleet, and then in 1588 were in the front line of attack against the English fleet, forming part of the Castilian squadron of the Invincible Armada.

These galleons had returned after the Invincible's disastrous encounter with English fireships and storms. The engagement had shown the greater manoeuvrability of the English ships – John Hawkins had removed the bulky castles which caused rolling. Another lesson of the defeat was the importance of smaller fighting ships; lack of these had prevented the Spanish approaching the Flanders shallows where the Dutch flyboats were able to foil the intended liaison of the Armada with the Duke of Parma's Army of Flanders. Consequently much smaller and faster vessels were designed and put into service.[74] Another inferiority of Spanish ships was evident in 1590 when the king sent Colonel Stadley, an English (Or Irish) Catholic, to El Ferrol to inspect the Armada and suggest ways of improving the installation of artillery. Stadley told Alonso de Bazán, commander of the fleet, that if Elizabeth's ships were to be matched six pieces of artillery would have to be placed at the prow and another six at the stern of each vessel. Bazán replied that apart from the shortage of artillery the ships did not have the gun ports to deploy so many guns because most of them had been built for commerce.[75] That was not peculiar to Spain; the English fleet which had faced the Invincible had been largely composed of merchant ships requisitioned by Elizabeth.

Technological conservatism is probably an inappropriate description of shipbuilding in Philip II's Spain. The English and Dutch introduced innovations and so did the Spanish; but there was no revolution in ship design anywhere in the sixteenth century comparable to the great changes of the fifteenth and nineteenth centuries. Shipbuilding has always been expensive and Philip II's indebtedness was a principal cause of delay and reduced output. Barros's frustrations at Guarnizo were to be experienced by others who struggled on the northern coast to repair the losses of the Invincible. It was the same story. The king urged construction 'with all speed' and 'saving all the hours you can' but soon there came desperate appeals from those in charge of construction that 'there is no money to buy a nail'.[76]

## 2 Artillery and munitions

The difficulties experienced by Barros in the 1580s in finding artillery for the new royal galleons were not unusual. Throughout the reign there were shortages: at the time of the Granada uprising as well as during the arming of the Invincible. The demands had become considerable; when the Invincible at last set sail, it had been equipped one way or another with 2431 pieces of artillery. The poor state of the royal treasury again had its effects. In July 1587, when the Invincible was being prepared, the captain-general of artillery, the royal official who supervised its manufacture, informed the king that because the promised 5000 ducats had not reached Lisbon, the construction of furnaces for the new foundry there was 'going very slowly'.[77] But lack of money was not the only obstacle. There was a shortage of technical skill at all stages of production. In 1575 the Council of War were investigating why the foundry at Málaga had failed to produce the 150 pieces of artillery requested even though the necessary 5000 ducats had been supplied; the explanation was that employment had been given to a founder who had elsewhere been dismissed for ineptness.[78] So acute was the shortage of founders that the captain-general was himself prepared to recommend the reemployment at Lisbon of Juan de Vallejo who had earlier been dismissed for unsatisfactory performance at the Málaga foundry.[79]

One way out of these difficulties was to import foreign technicians and artillery. The king did both regularly. He bought artillery from Germany and Flanders, and occasionally tried to borrow some from the duke of Savoy; other pieces were captured from enemy vessels. For the Azores campaign (1583) 26 pieces of cast-iron artillery were purchased from England, and three years later 24 more pieces were seized from English ships. In contracts with naval officers who supplied ships equipped for war, the terms commonly granted booty to the officers and crew; but all captured artillery was to be sent to the royal magazines. As in previous reigns the foreign founders continued to be hired in Germany or Flanders where, according to a Spanish military engineer, the craftsmen had the longest experience, greater patience than Spaniards or Italians, and cast guns in the best clay moulds which had been 'dried for years in the air'.[80] When the foreigners arrived they often complained of their treatment, like the Flemish working at Málaga in 1558–9 who were disappointed with their pay. Sometimes an imported technician failed to satisfy, as at Lisbon where Bartolomeo de Somariva, a Genoese, was said to be skilled and diligent 'but it is necessary to watch him; his foundings do not succeed'.[81] The king preferred to bring founders over from Germany to hiring his Milanese subjects to avoid weakening the Duchy's foundries. But the religious beliefs of German founders had to be carefully investigated; Philip did not want to pollute Spain with Protestants of questionable loyalty, however great their technical skill.[82] The king and his ministers saw the foreign technicians as a temporary necessity, hoped that Castilians would learn by working alongside them, and that eventually Spain would become independent of foreign skill.

Some of the artillery in use was made of cast-iron and occasional attempts to introduce this process in Spain had come to nothing.[83] But Spain and her enemies preferred bronze, because iron guns were soon weakened by rust. When Giovanni Rodi, an Italian living in Madrid, offered to establish the manufacture of cast-iron artillery in the Peninsula, the secretary of war (land) and the Council of the Indies tried to interest him in bronze instead; even though it cost more, the artillery would be more durable.[84]

Bronze required tin and much larger quantities of copper. Tests in Brussels commissioned by Charles V in 1521 had recommended 8 per cent tin and 92 per cent copper as the optimum alloy for guns, and proportions close to this were used in Philip II's reign. In 1554–8 36 bronze cannon were founded in Málaga for galleys at Barcelona; for this over 8100 *quintales* of copper and hundreds of *quintales* of tin had been imported from Flanders. When twenty years later Francés de Alava became captain-general of artillery he consulted factors 'to find out where the best copper comes from and at the most moderate price'; he was told that the best source was Danzig, the outlet for Hungarian copper.[85] But soon he was complaining that the prices quoted for Hungarian copper were so excessive and the terms so unfavourable that 'no business was done and the magazines are empty'; consequently he was unable to satisfy demands for artillery for Spain's fortresses. And he urged the king to consider investing in the mines of Cuba, where he had just learned that there was abundant copper ore. A former governor of Havana, Gabriel de Montalbo, had told him that Cuban copper could reach Seville at 3 ducats a *quintal*; Alava responded that even if it was double this price 'it would still be very cheap considering that copper from Hungary costs at least 14'.[86]

This was not the first time news of Cuban copper had reached Spain. Reports of its abundance in the area around Santiago had been received by Charles V, who sent out a German, Johann Tetzel, to the island in 1547 with a contract to extract the metal from the ore and reveal 'the secret' of the process to the inhabitants, which it soon became apparent he was reluctant to do.[87] After Tetzel's death Philip renewed the contract in 1578 with Sancho de Medina, a magistrate of Seville, offering him considerable additional concessions. By the terms of the agreement Sancho was required 'to find out the secret of operating the said copper mines' within two years of his arrival; and to take with him a workforce of 30 unmarried craftsmen and 100 married labourers with their wives and children. He would pay the king a duty of one twentieth of the value of the copper extracted over a period of ten years. In exchange Sancho was authorised to take possession of all foundry houses, engines, tools and other equipment used by Tetzel; he was also granted ownership of ten copper mines. He was given license to purchase 500 slaves and to cut wood for the mines without payment, provided no injury was done to a third party. He and his workers would be free to travel between Cuba and Seville without cost and for six years they would be exempt from the *almojarifazgo*, the customs duty on trade between Seville and the Indies. An additional incentive, licence to take land for housing and farming, was offered to the entire workforce. And the king gave Sancho the office of *regidor* of the town of Bayamo, some 50 miles from Santiago.[88] Philip was clearly anxious to encourage copper mining in Cuba, but while the contract specified Sancho's right to take copper to all parts of the Indies free of customs duties, there was no reference to exporting the metal to Spain and no mention at all of artillery. Perhaps the king's only interest had been income from the duty, though he may also have considered local production of artillery for the defence of strategic sites in the Indies. Whatever the king's motives had been Alava brought home the potential value of the mines for supplying all of Spain's artillery needs.[86] Alava wanted Sancho to operate the mines for the crown. He recommended that two master founders, skilled in smelting and refining, be sent to

Cuba along with six assistants 'so that in the course of time if these masters die there shall be someone who knows how to perform the founding'. And he asked for two other Spaniards to be sent out who knew how to make the charcoal needed as fuel for the extraction process. A sample of ore sent for assay to Seville by the governor of the island had been used by Alava to make a small piece of artillery. He was delighted with the result and was confident that in a few years it would no longer be necessary to buy Hungarian copper; Cuba would supply Spain with all that it needed and even with a surplus for export.

Alava's proposals were considered by the Council of War, which advised the king to adopt them, recommending that instructions be sent to the *Casa de la Contratación* to assist the scheme, particularly by requiring all ships returning from the Indies to bring copper, using it as ballast instead of the usual stone and sand. The king decided to implement the plan with the proviso, suggested by a minister, that first larger samples of ore be sent to Spain to ensure that the metal could be worked in large quantities as well as small.[89] Twenty years later the plans had come to nothing because, according to the Council of the Indies, it had not been possible to find craftsmen who knew how to smelt the ore. But now expectations were revived because Sánchez de Moya, a skilled founder, was in Cuba. The Council advised the king to establish foundries 'well inland' to protect them from corsair raids, and to ask Acuña Vela, captain-general of artillery, to provide Sánchez de Moya with models and specifications of artillery required. Confident hopes were again expressed that Spain would be freed from European imports of copper and that artillery would be produced in the Indies superior to what was manufactured in Málaga or Lisbon.[90] The king ordered the scheme to be introduced immediately. Artillery was founded in Havana and shipped to Spain; but by 1607 Philip III had brought this to a halt, preferring to bring the copper to Spain. This was after allegations that the copper was inadequately refined; it contained large quantitites of iron as an impurity, causing the guns to be weak. The furnaces were said to last no more than one or two foundings, while in Spain they withstood 15 or 20. And the operation had become so expensive that the copper was costing around 50 ducats a *quintal*.[91] The men sent to Cuba had been unable to find the secret of extracting and refining the metal. But the reports of the mines' richness had not been mistaken; in the early nineteenth century Santiago de Cuba was to become the world's leading producer of copper.

Projects were also conceived on several occasions during Philip II's reign to produce copper within the Peninsula. When a letter reached Milan with the news that a doctor in Venice was 'converting iron into copper' and making bronze artillery for the arsenal, it seemed that the correspondent was expecting alchemy to transform Spain's extensive iron deposits into copper. He was apparently unaware of the existence of copper mines in Spain, whose wealth, especially at Tharsis, had once amazed the Romans.[92] Philip had encouraged officials to look for mines within their jurisdiction. In October 1562 Perez de Mescua, *regidor* of Guadix, responded with an optimistic report of rich copper deposits in the Sierra de Baza and Alpujarras.[93] An assayer, Hernando de Velasco, had performed tests on the ore using a furnace erected in the *regidor*'s house, and concluded that 'the tests suggest that a large quantity of artillery of a very fine bronze could be made.' The local resources seemed adequate: timber, water, esparto for ropes and baskets, and there was a road from the mines to the coast. But the estimated cost of production – 15 ducats a *quintal* – was higher than the price of Hungarian copper and probably explains why no further action was taken here.

Copper mines were also discovered at this time at Madridejos (40 miles south-east of Toledo), but for months afterwards the two men involved had been unable to begin operations because they could find 'no one in the kingdom who knew how to fuse that metal'.[94] There were other difficulties in exploiting the mines of the Sierra Morena where the richest deposits were thought to exist. Since the beginning of the reign individuals had attempted to extract copper at Aroche and, not far from the Rio Tinto, at Aracena. Their difficulties were revealed when they gave evidence under oath to the officials of the *Casa de la Contratación* during an enquiry to determine if the mines could supply copper for artillery.[95] The *Casa* had the responsibility of supervising the manufacture of artillery for the Indies and for ships sailing there. Bartolomé Morel, a founder of bronze artillery and bells in Seville, testified that his father, Juan Morel, had begun to extract large quantities of copper in the area in 1558 but was soon 'impeded by disputes with individuals who laid claims' to the mine and so tried elsewhere, but this time 'the expense was greater than the gain' and so he abandoned that mine also. Juan began to operate a third mine but once again his work was brought to a halt by persons who laid claims. Eventually he lost most of his wealth. A similar tale of ruin was related by Rodríguez Tapia, though he still entertained hopes of profit. From these witnesses, the officials of the *Casa* heard that not a single copper mine was being worked in the entire Sierra Morena and that no guarantees could be given that copper could be produced there at a price lower than that imported from Hungary. Nothing more is recorded of this inquiry and copper imports continued. The crown intervened again in the early 1580s when the ambitious Enterprise of England began to take shape and quickened the search for additional strategic materials and military equipment. The king had brought over Nicolao Cipriano, master of the mint in Milan, to negotiate contracts for the supply of copper (and iron shot) from Spanish mines at half the imported price. Cipriano began a preliminary inspection and soon reported that there were many promising copper mines not only in the Sierra Morena but in Valencia and the Aragonese Pyrenees. Shafts would need to be sunk, and local timber and water supplies considered, to see which would provide the greatest profit. To do this he asked for 3000 ducats to carry out a systematic survey with 25 men working for a year. The assessment of this proposal was delayed by the death of Alava, captain-general of artillery, and the absence of certain ministers; and Cipriano was threatening to return to Italy because he had spent 2000 ducats of his own money while he was being kept waiting for his salary. In March 1587 the Council of War advised the king that this was a project of 'great importance,' that Cipriano be given what he requested, and that he be asked to survey tin as well as copper mines. The king authorised these recommendations and the survey began. On Cipriano's advice 8000 ducats for the eventual operation of the mines was obtained by selling the title of 'duke' to 'a titled person of noble blood and suitable income'; the count of Trevento (Naples) was willing to oblige, but the whole project was halted by Cipriano's illness and death. Cipriano's promise to save the king over 200 000 ducats by eliminating the import of copper and shot vanished.[96] None of the bronze artillery for the Invincible Armada was made from Spanish copper or tin (see also plate 3).[97]

The chronic shortage of native technical skill was evident in 1590 when the captain-general of artillery told the king that although

Cipriano had discovered copper mines close to the foundry at Málaga, none had been extracted because 'there was no one who knows how to do the fusion'.[98] The economic difficulties of copper-mining at this time were indicated by Karl Schedler, a German whom Philip appointed at the end of 1594 royal administrator-general of mines in the Peninsula. His duties included encouraging Spaniards to involve themselves in mining enterprises. Accordingly he at once asked the king for financial incentives, alleging that men were deterred by the heavy initial expenditure required for such risky commercial ventures, and by the rising costs of labour. And on top of this they had to pay duty to the king for every mine worked. Schedler therefore recommended the introduction of a system of royal loans to offset the initial expenses, and a reduction or delay in imposing the royal duty.[99] Philip's response is revealed in his instructions to Gerónimo de Ayanz who was soon appointed administrator-general after Schedler's death. The king was unwilling to forego the income from mining duties: if the men who discovered mines were too poor to pay the duty then Ayanz must find wealthy men willing to come to an arrangement with impoverished owners whereby the duty would be paid in exchange for a share in the profits. Philip's soaring debts had forced him once again to suspend payments (1596) and he was clearly in no mood to consider loans for mining ventures or to reduce his potential income from them. He urged Ayanz to do his utmost to activate copper mining. He recognised the difficulties of rising wages; and the problems created by workers who returned to the fields at harvest time, causing untended mines to become flooded, and requiring 'even greater costs' to resume operations than had been needed at the outset. But in the end the king had nothing to offer except that Ayanz consider the desirability of importing Germans 'so that mines might be operated with more industry and less cost and to train natives of these kingdoms'.[100]

Spain's failure to exploit its rich copper resources was due to economic and technological deficiencies and not to any lack of government interest. The particular metallurgical difficulties are not easy to identify because the sources give little indication of the processes that were being attempted. But the frequent references to 'fusion' show that the difficult smelting of copper ore was being attempted and not the simpler 'wet process' of forming a soluble copper salt and then precipitating metallic copper by addition of

[G.A. - 199

Hanse de haber para la Isla de Canaria quatro culebrinas como la que aqui va [illegible]
que tire diez y ocho libras de pelota cada una y ha de tener de largo treynta diametros /

La reparticion de los miembros que han de llevar ha de ser
en la culata tres diametros y un sexto; en los muño-
nes dos diametros y cinco ochavos; en el cuello dos diametros; y las cintas y las cintas y molduras [illegible] muño-
nes y culata se ha de guardar la orden y forma que va
en el dibuxo de la dicha pieça, teniendo advertencia
que las molduras que hubieren de entrar dentro del encaval-
gamiento no levanten demasiado porque no se a menes-
ter tomar la madera [illegible] [illegible] [illegible] de [illegible]
ancho que sera [illegible] /

Pareçe que esta pieça pesara [illegible] quintales de Castilla.
Hanse de poner en ella las armas Reales muy bien
hechas relevadas con la corona, y debaxo del escudo
una forma de rotulo [illegible] y dentro unas letras de
relieve que digan desta manera: Don Phelippe
y Rey de Spaña. Y en la cinta que anda debaxo
del dicho rotulo unas letras labradas [illegible] con
buril que digan desta manera: Don Joan de
Acuña [illegible] Capitan General de la Artilleria
1587 años. y no ha de llevar otro ningun
letrero /

El cobre de que se han de haber estas culebrinas ha
de ser de Ungria de tableros quadrados [illegible]
[illegible] [illegible] de [illegible] de la liga de estaño [illegible]
[illegible] a valor de ocho por ciento [illegible] de Castilla [illegible]
otro ningun genero de metal ni [illegible] [illegible]
do de comprarlo al mas aventajado preçio que
se pudiere /

Ha de haber estas quatro culebrinas [illegible]
[illegible] fundidor de Su Magestad en Sevilla y por [illegible]
y por la [illegible] su mano [illegible] y [illegible]
se le ha de dar [illegible] lo mesmo que Su Magestad
[illegible] por quintal de lo que [illegible]
y por su cuenta se funde y en la prueba [illegible]
della se ha de tener [illegible] que con las que se
hazen para Su Magestad y lo mismo en lo de los encavalgamentos.

Pareçe que cada una destas culebrinas llegara a
ochocientos ducados castellanos.

De manera que las dichas quatro culebrinas costaran
tres mill y doscientos ducados [illegible] [illegible]
los quatro mill [illegible] que Su Magestad [illegible]
[illegible] aquella isla [illegible] ducados que [illegible]
[illegible] en ellos para los quatro encavalgamentos
guarnecidos para cada una de las dichas culebrinas
de [illegible] costando cada uno ciento y cinquenta ducados
y [illegible] han de haber los maestros de carpinteria
y herreria que Su Magestad tiene en Sevilla y [illegible]
[illegible] [illegible] [illegible] [illegible] [illegible]
[illegible] la misma razon y como se les [illegible]
en las obras de Su Magestad /

Las dichas pieças se han de haber acabar y perfecio-
nar a contento y satisfacion de Don Juan
Duarte factor y Justicia de la casa de la contra-
tacion de Sevilla por Su Magestad. Fecha en
Lisboa a veynte y cinco de Julio 1587.

Don Juan de Acuña

iron. It is also clear that the separation of iron, the principal impurity in the extraction of copper from its ores, was causing difficulties.

The failure does not indicate that Spain was peculiarly backward in technology and enterprise. In England similar attempts had been no more successful. There too Elizabeth had tried to promote copper mining with a view to securing independent supplies of bronze artillery. And there too foreign technicians and administrators had been brought in to assist. But little copper was produced and soon the mines of Cumberland and Cornwall had ceased to operate, forcing England to rely on imports in the seventeenth century.[101]

The same combination of economic and technological difficulties help to explain Philip's constant shortage of gunpowder. Saltpetre, the principal ingredient of the explosive mixture, was abundant in the earth, and in animal excreta in stables and pigsties; yet throughout the reign supplies at the royal munitions houses ran low. Sulphur, another ingredient, had been produced in Roman Spain. The existence of deposits at Hellín (Murcia) was known from at least 1562; yet stocks were always scarce and the crown relied on imports from Italy. Charcoal, the third ingredient, alone seems to have presented no problem. At times of particular urgency, the gunpowder itself was hastily imported, as in 1587 during the equipping of the Invincible when much of the Armada's 5175 *quintales* of powder had to be brought from Italy or purchased from German merchants.

The production and sale of gunpowder and saltpetre was a royal monopoly. The arrest in 1558 of Juan Vidal, maker of gunpowder in Lorca, for unauthorised activities indicates the abuse which the crown had sought to eliminate by its monopoly; he had bought up large stocks of saltpetre and sold them abroad, creating shortages and a rise in prices. In Seville in the 1570s action was taken to curb

Plate 3 Bronze culverin, one of four ordered in July 1587 for the defence of the Canaries. They were to be manufactured in Seville by Juan Morel, the king's founder, at an estimated cost of 900 ducats a piece. Weighing around 60 *quintales* and proudly embossed with the royal coat of arms and the names of the king and captain-general of artillery, the specifications of their alloy reveal Spain's continuing dependence on foreign imports of strategic material: the copper was to come from Hungary; the tin from England, at a time when the Invincible Armada was being planned.
(AGS:GA 199/52; *Mapas, Planos y Dibujos* V–16).

the profiteering of Raimond Martin, a Frenchman licensed by the crown to make gunpowder for the armada. But he had employed Frenchmen in several localities of Andalucía to secure all available stocks of saltpetre in order to sell gunpowder at an excessive price to the crown and – it was alleged – to enemies of Spain.[102] High-quality gunpowder, free of impurities, was the other goal of the royal monopoly.

Controls on supplies were not easy to enforce. When the king learned that much saltpetre and gunpowder was being smuggled from Aragón to France via Navarre, he asked the viceroy of Navarre if a royal prohibition on the traffic would in any way conflict with the *fueros* of Navarre; the reply was that it would cause 'great offence'.[103] But even in Castile where there were no *fueros* to be observed the numerous infringements and repeated reenactments of the royal monopoly bear witness to its ineffective enforcement.

The *Priorazgo* (Priorship) de San Juan (New Castile) was the main centre for collecting and refining saltpetre. The crude saltpetre was digested in boiling water and then poured into vessels where it crystallised. This purification was not always done with sufficient care; as a result gunpowder made at Málaga's royal munitions house had to be reprocessed.[104] The state of the crown's finances halted saltpetre production at Tembleque and Lorca because the workers were unpaid. Production was also impeded by local disputes. In the *Priorazgo* firewood must have been short because esparto grass was used instead as a fuel for the refining process. But this conflicted with the needs of the inhabitants who used the grass for cattle grazing and as a fuel for cooking. The king assuaged local resentment by closing specific areas of land where esparto was growing to saltpetre workers for a period of five years.[105] Local disputes also occurred at Cartagena where in 1591 the justices were said to be hindering the collection of saltpetre earth [106] – perhaps, as in England, the privacy of residents was being ignored.

The production of gunpowder at the royal munitions houses depended on sulphur supplies from Italy; but these did not always arrive, or there were worrying delays, as in August 1587, when the ships from Italy inconveniently unloaded their sulphur at Gibraltar instead of Málaga where the lieutenant-general of artillery was waiting anxiously to receive it.[107] The prospect of securing dependable supplies within the Peninsula and at less cost was therefore given serious examination by the crown. Eventually Philip was to invest more for this than for the copper mines and, because the technological demands of sulphur extraction were much less than those of copper metallurgy, some success was achieved.

The king had received a report from the superintendent of gunpowder manufacture at Tembleque that potentially important sulphur mines were being worked by a dozen men near the río Segura. That was in 1572 and although the king at once ordered an investigation nothing came of it.[108] But positive results came later in 1587 when mines in the same area were offered for sale by Francisco Monreal of Murcia. His father had discovered mines at Hellín and Moratalla in the 1560s and had been given permission to operate them for his own profit in exchange for the usual royal duty. But his wealth had been spent on bringing 'masters from various parts' to teach him 'the secret of the fusion,' and when he died he was destitute and still without the secret. And the son did not want to suffer the same fate through the mines which he had inherited. He had travelled to Italy to observe manufacture there, taking with him a *quintal* of sulphurstone from his mine. In Italy, as at Hellín, the sulphur deposits were native (that is uncombined) and required the simple process of melting, the sulphur flowing off from earthy mixtures. Monreal claimed that after 'great expense' he had returned to Spain in possession of the secret. For the secret and the mines he was now asking 60 000 ducats.

Hernando de Acosta, lieutenant-general of artillery at Cartagena, was sent to inspect the mines and report on their 'quality and quantity'. He spoke to men who had worked in them and consulted a local *zahorí*, whose occult powers of detecting subterranean objects strengthened the belief that the mines were 'so rich that they will not be exhausted for a very long time'. Acosta was impressed with the timber and water resources, and with the local soil which he said would provide strong material for the buildings of the plant. He estimated that a *quintal* of sulphur would arrive at Cartagena at 12 *reales*, less than half of the cost of Italian imports and much less than the 100 *reales* paid to merchants in Burgos the year before.

Acosta's report went to the Council of War in April 1587. It was warmly received; the Council valued the existence of 'mines of such quality in these kingdoms and in such a good locality' and recommended their purchase by the crown. Monreal demonstrated the process of extraction to officials in Madrid and to the king in the Escorial, producing a yield of about one quarter of molten sulphur from the crude sulphurstone. But the king was slow to act, causing the Council of War a year later to request a quick decision so that gunpowder production was maintained 'in these present dangers'. Monreal complained that he had been kept waiting for months in Madrid at great expense to himself. The sale was at last agreed in May 1589. Monreal got less than he had asked – 20 000 ducats: 9000 for the mines and 11 000 for the secret with the added condition that the yield of pure sulphur would be as Monreal had guaranteed. He would stay on as an adviser, teaching the process to employees.[109]

It soon became evident that the king's purchase was not as advantageous as it had first appeared. When the captain-general of artillery went to the site he found that the mine was in a deep ravine liable to blockage by falling earth and rocks, and that transport of sulphur from the mine by cart would not always be possible because the track would be destroyed whenever there was heavy rain; instead pack-animals would have to be used. Worse, the site was remote; the nearest populated locality was two leagues away and could be reached only by a poor road and a difficult river crossing – sometimes the strong currents were impassable for a whole week. Therefore the workforce would have to live on the site and special arrangements made for provisions.[110] But who would want to work there? The administrators of the mine had offered work to craftsmen for miles around, in Hellín, Caravaca and Murcia, but reported that 'so far no one has come here to negotiate'.[111] The suggestion that towns in the area deliver criminals to serve in the mines for periods of two to four months was rejected.[112] Forced labour was however used to tile buildings of the plant; but the local justices were said to be 'reluctant to act against the residents' and so the administrators asked the king to authorise 'the compulsion of all workers and labourers needed, paying a just rate' and 'the requisitioning of wheat, barley, rye, wine, bacon, cheese, oil and vegetables needed to feed the men'.[113] It is not clear if these measures were approved.

There is also evidence that the extraction process was running into difficulties. Alonso Carrasco de Cuellar, who had been appointed to supervise the operation, thought the process was simple enough in principle: 'there is not much science to it and it can be easily learned by anyone with a rational mind'. Nevertheless he found it was not so easy to put into practice. The first furnace used was inefficient: fuelled by wood fed into two apertures, it took a long time to heat up and the fusion of the sulphurstone was slow, taking 30 hours and consuming so much wood that fears grew that the surrounding forests would soon be exhausted. Ventilation was inadequate and sulphurous fumes so affected 'the eyes, throats and chests' of workers that 'none will dare to return to the furnace until this is rectified'. Carrasco asked that the king write to ministers in Italy to communicate information and sketches on furnace design, ventilation and receivers, and for Italians experienced in the process to be brought over. The response of Andrés de Prada, secretary of war (land), was understandable: Monreal said he had studied these things in Italy; let him provide the remedies.[114] By January 1591 the furnaces had been shortened, chimneys provided and the process was now 'four times faster than at first.' In eight fusions 664 *arrobas* of sulphurstone had yielded 120 *arrobas* of sulphur. Carrasco was busy with experiments to improve receivers, many of which had broken, bringing work to a halt. He was also testing materials to find a more durable furnace, and thought he had recognised in a local stone the same material which he had seen used for the founding of artillery in Málaga and Lisbon because it 'withstood the fire better'. He was going to make 'a small cup' of the material and test its durability in a furnace.[115]

All of this suggests that there was no shortage of initiative at Hellín. But the output was so disappointing that in July 1597 the king asked Ayanz, the administrator-general of mines, to find out if there were sulphur mines elsewhere in the Peninsula which could be run more cheaply. And suspecting that Monreal had broken the contract by failing to produce the promised yield, he appointed an official to investigate.[116] But Monreal remained an administrator at Hellín until his death in 1610 and the mines continued to be operated for the crown in the seventeenth century. By the 1620s Hellín was able to supply the munitions houses with substantial quantities of sulphur; but it was still not clear that the quantity was enough to make Spain independent of foreign supplies – Philip II's intention – and to justify a ban on imports.[117]

The ingredients of gunpowder were mixed together and ground into granules at Málaga and Cartagena. Shortage of money at Málaga in 1586 caused production there to fall from 8 to 3 *quintales* a day.[118] But the king's plans to double production at Málaga by using 'larger mills and engines' on an extended site met municipal opposition; the city protested that a larger site would bring production dangerously close to tall houses, and any spark escaping from their chimneys or hearths would be enough to cause a conflagration.[119] In another attempt to boost production a privilege is said to have been granted in 1594 to Francisco Trujillo, powder maker of Granada, for an improved engine which its inventor claimed would reduce the required workforce to one tenth.[120] But in the same year Philip was seeking extra workers 'to make a larger quantity of gunpowder in less time': he instructed the Inquisition to remove impediments to the employment of Moriscos in Aragón for this purpose.[121] And this despite widespread suspicion of their loyalties. When a huge gunpowder explosion shook Seville on a May afternoon in 1579 destroying many houses and killing 100, the immediate reaction of the *asistente* was that it had been the evil work of a Morisco; but he soon concluded that the most likely cause was a spark flying from a mortar in which charcoal was being ground.[122]

Apart from the shortages of gunpowder and artillery, cannon balls were often not available. Vizcaya may have had the richest haematite deposits in Europe and an established armaments industry supplying large quantities of pikes, arquebuses and muskets from its numerous water-powered forges;[123] but iron cannon-balls were not produced there. In July 1572 the king wrote urgently to Italy for two masters to be sent to Madrid 'as quickly as possible' because of the 'great shortage in these kingdoms of cannon-balls' and 'there is no one here who knows how to make them'.[124] The cannon-balls were usually imported from the Netherlands or Milan. In October 1574 the king was seeking the cooperation of his ally, the duke of Florence, to exploit the iron on Elba and make cannon-ball there 'all of the same calibre so that they can be used for all pieces of the same type'.[125] But the shortages continued. In July 1586 the king was informed that 'most of the artillery of Fuenterrabía, San Sebastián and Perpiñán have not a single shot to fire' and in the following year the Council of War was pressing the king to act because the Enterprise of England was threatened by lack of shot.[126] Eventually the 123 790 cannon-balls taken by the Invincible had to be imported.

At Euguí near Pamplona there was a royal ironworks specially intended to supply cannon-ball. It was in a state of disrepair and needed improved moulds for making the shot. Although it was reactivated in 1589 it produced little in Philip II's reign.[127]

### *3 Gunners and engineers*

Men as well as strategic materials had to be imported to maintain Spain's military power. The king and his ministers wanted to end Spain's dependence on foreign gunners and military engineers, and for this purpose they considered incentives to encourage men to take up these professions and created technical schools to provide the necessary training.

It was never easy for the crown to find enough gunners for its armadas and fortresses. The occupation was hazardous even in time of peace. Gaspar de Morales, master gunner of Mallorca, lost two of his sons who had become gunners: one was captured in Valencia and taken to Algiers where he died; the other was blown to pieces during the testing of a new piece of artillery in Lisbon. In a similar accident during the testing of artillery at Lisbon Anton Pollo, captain of the Sevillian gunners, lost an arm and two other gunners were killed.[128] The prospect of being ordered to serve in the harsh conditions of the African *presidios* was certainly a deterrent, and Alava suggested that a penalty of four years in the galleys might be introduced to prevent qualified gunners from going into hiding when called to serve.[129] Irregularity of pay, a constant grievance in Philip's armed forces, also affected the gunners. At Cartagena the captain-general of artillery thought six gunners were needed but said 'no one wants to fill the vacant positions' because the money for their wages, supposed to come from fines to the treasury, was being used for other purposes by the *corregidores* of Murcia.[130]

Artillery schools already existed at Burgos, Barcelona and Gibraltar but they had not satisfied the growing demands. In 1575 the king instructed the Council of the Indies to create an artillery school at the *Casa de la Contratación* in Seville to supply gunners for the Indies fleets. Alava hoped that this would get rid of the Flemish, German and French gunners who, after acquiring experience of the Indies passage in the service of Spain, transferred to ships of their own nations in order 'to rob the Indies and Spanish shipping'. Andrés de Espinosa was appointed to supervise the school, with instructions to ensure that those who entered were true Christians over the age of 20, had good eyesight and sound limbs, and had experience of at least one voyage to the Indies. The basis of their training was a period of two months' instruction in making powder and incendiary mixtures, and firing artillery in the presence of professional gunners. When this was satisfactorily completed the candidate was given an oral examination with questions relating to pieces of artillery and various gunners' instruments which were shown to him. Some of the questions asked were: 'What is bronze made from and what is the best alloy?' 'Why do artillery pieces burst?' 'How are smaller and greater ranges of fire achieved from a given piece of artillery?' 'If I find myself in some part of the Indies with powder and lead but no artillery how could I defend myself from the enemy?' 'If you were in a fortress how would you know if it was being mined and in what part?' The answer to this last question was: 'Put a little mercury in a basin and take it to wherever mining is suspected; if mining is in progress the mercury will tremble'. If the candidate answered well he was issued with a certificate by the officials of the *Casa de la Contratación*, licensing him to serve as a gunner on any ship that sailed to the Indies.[131] The school functioned until Espinosa's death in 1592 when, after amalgamation with another class, the formal training of gunners in Seville continued until 1681.

The instruction at Seville was predominantly practical. Espinosa expected no more knowledge of range from his students than that firing at point blank hit the target at close range, and an elevation of 45 ° achieved maximum range. But others attached greater importance to teaching gunners mathematics. The king is said to have wanted a chair established at Salamanca precisely for this purpose.[132] Spanish treatises on artillery by Luis Collado (1592) and later by Diego Ufano (1613) included extended discussions of the theory of ballistics.[133] But when Giuliano Firuffino came from Milan in 1585 to teach Spanish gunners at the school of Burgos and later, during the 1590s, in Seville, his mathematical approach was criticised by men experienced in the art of gunnery. Soon after Firuffino's arrival at the munitions house of Málaga, where he had also come to teach mathematics to gunners, Juan de Acuña Vela, captain-general of artillery, wrote to the king that Firuffino himself would benefit greatly from his stay there because 'he has discovered that while it may be useful to learn the art of artillery through science, this is best done manually; and so he has been freed from the illusions he held concerning the certainties of mathematics'.[134] Tensions existed between the theorists and the practitioners. It is clear that all attempts at mathematically precise theories of ballistics or tables of range were irrelevant to the crude guns of the time. In all parts of Europe guns were inaccurately bored; their bores were also purposely widened to prevent blockage by the irregularly cast shot. With shot leaving cannon at unpredictable angles and with unpredictable force because of the wide variations in gunpowder, trial and error was a much better guide than mathematics. Accurate firing at long range was not possible; nor was it desirable to attempt it – the greater powder charges would have increased the likelihood of bursting the weak guns. Consequently sixteenth-century gunners from Spain and elsewhere fired at close range and without assistance from mathematics.

There is clear evidence in the 1590s that existing artillery schools were failing to produce gunners in the required numbers. The duke of Medina Sidonia had heard from the king 'how little was the interest among the naturals of this kingdom to become gunners' and advised the creation of numerous schools all along the coast with grants of privileges to those who entered. Eventually the king had to ask the German princes for gunners: 140 of them, the majority having experience of war against the Turks in Hungary, arrived at Puerto de Santa María. They were soon clamouring for pay, and recommendations were made for their removal inland to prevent their escape.[135] In May 1595 the king granted gunners qualifying at Seville some of the privileges associated with nobility: licence to carry arms; exemption from the billeting of soldiers in their houses; freedom from arrest; also rights to trial in the special court of the *Casa de la Contratación* and, when at sea, by the officers of the armadas. But this bore no fruit in the long term. The continuing failure to give adequate pay – at Cartagena in 1587 it was no more than one *escudo* a month[130] – or to guarantee the receipt of any pay at all seems to have been a powerful deterrent to gunnery in Spain.

Engineers enjoyed much greater prestige and were paid handsomely for their work on civil and military projects. But they too were on occasions kept waiting for their salaries. In December 1575, three months after Philip's mounting debts had again led him to suspend payments, Giovanni-Battista Antonelli, a valued royal engineer, was about to leave Madrid for works in Barcelona when he was arrested for failing to pay a private debt of 6000 *reales*. His letter to the king from a prison in Madrid begged for the salary that was due to him.[136]

Fortification was the principal reason for Philip's eagerness to acquire engineers. The new Italian system with its bastions and regular polygonal trace had slowed the pace of sixteenth-century land warfare, keeping the enemy at a distance and eliminating the former superiority of artillery attack. The men who knew how to construct the impregnable defences were in great demand throughout Europe. In October 1561 Philip wrote to the duke of Savoy requesting the loan of Francesco Paciotto of Urbino, then working in Turin. He wanted to consult the engineer on fortifications, and urgently: 'the quicker you send him to me the happier I will be'.[137] Paciotto subsequently fortified La Coruña and designed the famous citadel at Antwerp. When Granvelle was viceroy of Naples, Philip asked him for a detailed report on engineers in the kingdom, their experience and their salaries, with a view to bringing them to Spain. Granvelle replied that none of the nine he listed had the necessary experience, though he singled out 'Jacobo, a Fleming' who was 'not very expert in fortification but is very well trained in mathematics and mechanics, has a fine intellect and so we can expect very good service from him'.[138]

News on engineers, indicating competence in fortification, current salary, and purity of blood, was also sent to the king from the embassies. There was a certain amount of intrigue in attempting to lure to Spain or parts of Philip's empire engineers employed in the service of other princes. From the embassy in Venice Philip learned that Orologio of Vicenza, an engineer 'who is reputed to have a better understanding of the art of fortification than anyone else in Italy' was discontent in spite of the high salary given him by the Signoria. Word had reached the Spanish embassy that he would be willing to serve Philip instead and this was confirmed in secret discussions with him. The embassy asked the king for a quick decision, emphasising that, apart from his familiarity with Venice, 'this man knows every corner of France'.[139] This soliciting of engineers could cause resentment when it resulted in the loss of a valued military expert. Venice protested to the Spanish ambassador when Giovanni-Paulo Ferrari left to serve Philip in Naples.[140] Philip was sometimes the loser. In 1572 he heard from his envoy in Turin that an Italian engineer who had recently completed fortifications in Flanders had become disgruntled and switched his services to France. There was concern about what this man might do for the enemy and Philip commented that 'it would be desirable to know his name' and inform the duke of Alva, captain-general in the Netherlands.[141]

Within the Peninsula Spanish military engineers were rare. When the captain-general of artillery was asked to recommend an engineer to design two forts to defend the Straits of Magellan from Drake's raids he said he knew of seven engineers in royal service in the Peninsula and 'all are foreigners. I know of no Spaniard who is equal to them'.[142] The foreigners were Italians. At least 25 of them can be identified as having worked in Spain during Philip II's reign. They strengthened defences in the Balearics, on the border with France and along the Mediterranean coasts (plate 4). They built elaborate forts in the Indies and Africa. And they were predominant in the siege warfare in the Netherlands.

When Giovanni-Battista Antonelli came from the Papal States in 1559 to inspect the defences of the kingdom of Valencia, it was the beginning of a family tradition. His brother and nephew soon followed and the Antonellis were continuously in the service of the Spanish monarch up to the middle of the seventeenth century.[143]

The lack of Spanish military engineers was not far from Philip's thoughts on Christmas Day 1582, when he gave approval to the foundation of an academy of mathematical sciences in Madrid. He had wanted its curriculum to be imitated in schools throughout Castile, but the municipal authorities would not bear the cost.[144] But in Madrid the new institution, financed by the crown, continued to function from 1583 to 1625, offering lectures on military engineering, navigation and architecture. The professor of fortification received 375 000 *maravedís* a year, twice as much as the highest salary of any professor at the university of Valladolid, an indication of the king's esteem for the subject. No documents have yet been found which give precise information of the numbers who attended the Academy. A good deal is known about the content of the teaching because some of the lecture courses were published. Cristóbal de Rojas had advised on the fortifications of Pamplona and those of the Spanish expeditionary force in Brittany before taking up his teaching appointment at the Academy. His treatise

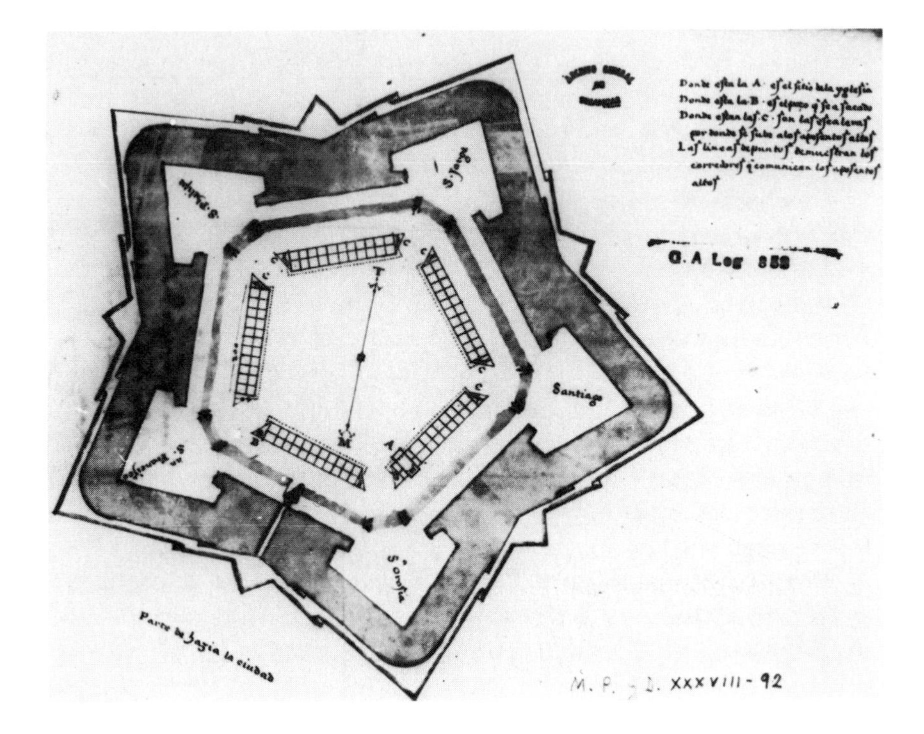


Plate 4 The citadel at Jaca (Aragón) constructed by Tiburzio Spanochi of Siena, one of several Italian military engineers in Philip's employ. The plan shows the polygonal bastioned trace characteristic of Italian military architecture. This fortress was built in 1592 soon after Philip had sent an army to quell disorder in Aragón, and was partly intended to prevent invading bands of French Protestants entering from Bearn. The fort included a church at A and a deep well with 'very good water' at B. Around 5000 ducats a week were spent on the construction, which Philip was eager to see completed.
(AGS:GA 353/3; *Mapas, Planos y Dibujos* XXXVIII–92).

on the theory and practice of fortification was a publication of his lectures, intended to supply 'the Spanish with everything they needed for war'. In the prologue he described the success of his course, perhaps with exaggeration: 'Many virtuous soldiers attended the Royal Academy and the effects were great, because after a few lessons they began to prepare plans of fortification with such proportion and reason as if they had been doing it for many years'.[145] Rojas told his students that fortification required three skills: knowledge of 'much mathematics', especially Euclid because this provided the basis of the theory of proportions on which all construction depended; arithmetic 'for calculating the costs of construction and many other things'; and the ability to survey sites for fortresses. He was convinced that soldiers ignorant of mathematics would be in no position to build fortifications; but also advised that geometricians with no experience of war should be accompanied by veterans when selecting a site for defence. The indispensability of geometry for fortification had become a commonplace in earlier Italian treatises.[146] And Rojas was similarly following Italian practice when he gave rules for constructing regular polygonal fortresses with up to seven bastions.

## *4 Secret inventions for war*

Numerous military inventions, mostly from Italians, were offered to the king, usually in the hope of financial reward and on one occasion in an effort to secure release from a royal prison.[146a] From the Spanish embassies in Milan and Venice came offers of rapidly firing artillery; accurate guns which 'always hit the bull's eye'; more mobile artillery; transportable bridges; amphibious vessels; methods of preventing rust on arms; portable mills for grinding wheat in fortresses under siege. The inventors often described their secrets as 'never before seen' or 'never offered to any other prince'. As a check on their novelty Philip was advised to keep in the Escorial library a book describing numerous engines of war by the deceased cosmographer Alonso de Santa Cruz.[147] The inventions were not always concerned with the weapons of modern warfare. When Giovanni Bianchi of Crema (Milan) offered Philip 'mirrors, in the form of shields, to dazzle the enemy and burn their clothing or ships' he reminded the king that this was the stratagem used by Archimedes to burn the Roman fleet at Syracuse.[148] It was a sign of the considerable contemporary interest in the classical tradition of warfare.

Some of the inventors were veteran campaigners like Andrea Dragovic of Cattaro who had fought against the Turk in Hungary. The ambassador in Venice was impressed by his bearing: 'he seems to be a sensible, modest and Christian man, and he will be content with a fair reward for his secret'. The ambassador was also impressed by the demonstration of the invention. This was a multiple-barrelled gun composed of a dozen arquebuses or muskets, arranged in rows in a case, supported on a tripod, and firing simultaneously at various heights; one man loaded and another fired. It was for use on ships or, with more effect, on land where four at a time could be carried in specially designed wagons. And the wagons, armed with spikes, joined together at their ends and sides to form what the inventor described as 'a marvellous stockade in which 10 000 infantry can gather, safe from any cavalry charge'. There is no record of the outcome of this offer, but Luis de Requesens, captain-general in the Netherlands, was very interested[149]. Another invention which attracted interest is shown in plate 5.

Several of the inventions unexpectedly came from the clergy. In February 1590 fray Angelo Meladao (a Portuguese?) was busy in the royal munitions house at Málaga struggling to complete his new type of artillery and, according to the Council of War, causing expense and delay by complaining too much.[150] At El Ferrol in October 1589 Martín Rico, a Jesuit, was demonstrating his artifice of wheels on a galleon, probably an attempt at propulsion without oars or sails; it failed because the galleon 'needs more wheels than can be fitted into it'.[151] And the invention for moving ships in calms considered at Málaga in 1590 had come from a Dominican monk.[152]

This involvement of the clergy in designing military devices must have seemed strange at the time, because when Gonzalo Venegas sent the king an account of his various inventions for war, he felt it necessary to apologise for 'meddling in affairs so alien to my profession'.[153] This ecclesiastic declared that he carried 'no other standard but the ensign of the cross and wore no other armour than the dalmatic'. He said his heart, like those of all Christians, had been deeply touched by the loss of La Goleta (1574) to the Infidel, and the war machines which he had been motivated to design were born of his 'great zeal to honour God and defend His law'.

Most of his devices were incendiary materials to set fire to the Turkish fleet. A lantern planted on the poop of the enemy commander's galley would explode when lit. But his most elaborate plan was to trick the Turk into taking on board their galleys sealed copper vessels containing 'ham and other food the Turks themselves

Plate 5 Defence against a cavalry charge. This trestle fitted with a rope net was invented by the Italian engineer Bartolomeo Campi for Philip's Army of Flanders. An assemblage of these units along the flank of a squadron of infantry would, it was believed, disrupt any attack of enemy cavalry, the horses stumbling as their feet became entangled with the nets. They were deployed at Maastricht.
[*Comentarios de Don Bernardino de Mendoza de lo sucedido en las guerras de los Paises Bajos desde el año de 1567 hasta el de 1577* (Madrid, 1592), p. 81. Reproduced with permission of the Bodleian Library: 4°A 45.Jur BS]

do not eat but give to their Christian captives.' The vessels would also contain a mixture of gunpowder and tar, so that when the enemy put the pots on a fire to cook, an explosion would occur scattering burning material and causing a conflagration, spread from one galley to another. How could the Turks be led into the trap? It would be necessary for a forged official correspondence to fall into their hands, purportedly containing advice from Christian generals of galleys that for cooking meals for seamen and soldiers it would henceforth be preferable to use sealed copper vessels, instead of the usual thick earthenware, because they would last longer. But the most unrealistic part of the plan was the assumption that all of this might be achieved without loss of life; it was the only thing that Venegas asked the king for in return: 'because I am of the clergy I am not permitted to enter into things which cause death or the spilling of human blood; and if these instruments are to be used it may only be for burning enemy galleys or fortresses and not to cause death or bloodshed'.

Inventors' projects were often referred to the Council of War whose members included military experts; if there seemed to be any prospect of success they would recommend demonstrations or tests in the presence of a named official. When Stefano Tirone of Picerno (Naples) offered the king a secret method of preventing the explosion of gunpowder stores by sabotage or by accidental lightning strikes, the Council of War was sceptical and inclined to the view that 'this invention like many others will turn out to be useless'; nevertheless it recommended tests to be performed in Milan in the presence of the governor, his captain-general of artillery and other experienced persons, if only to expose the invention for what it was.[154]

An unfavourable report was given by Francés de Alava after he was asked by the king to interview Giuseppe Bono, a Sicilian who had arrived in Madrid with promises to supply 'most terrible mines'; artillery capable of being dismantled so that 'enemy spies would not know what was being carried'; and a technique to make any fortress impregnable. Alava thought the inventions useless or lacking in originality, and as for Bono 'the worst thing about him is his excessive confidence'. The final decision on inventions as in everything else rested with the king. Philip's intervention was felt in the negotiations with Bono. Alava had spoken to Bono of his concern that there should be safe access for ships to Seville, then impossible because of the silt and sunken ships in the Guadalquivir, apart from the notorious bar. Alava was sceptical when Bono alleged that he had the means to clear the river; but Juan Delgado, secretary of war, advised the king to pay Bono 160 ducats for this task. Philip responded: 'it is no good sending him to Seville unless we are certain that he will do what he says because he could cause us much expense without any benefit'. He wanted further consultation.[155]

Inventors were commonly kept waiting for the king's decision. Martin Altmann, a German, complained to the king that he had waited three years and spent much of his money on models and apparatus for the construction of war engines. His list of inventions included hollow cannon-balls filled with venom, whose fumes would kill all who breathed them; and a rapid method of repairing ships holed by artillery fire. Philip believed there could be things of use here and at last agreed to employ him with a salary of 25 *escudos* a month; he also issued instructions to the captain-general of Portugal for 'a test to be made of each thing on land or sea and inform me of the outcome'.[156] Plate 6 shows another tested invention.

When the Council of War or other experts dismissed military inventions, that was usually the last that was heard of them. But even then the king might intervene to give the inventor another chance. This happened in February 1589 when Francisco de Nieba approached the king a second time with his offer of two inventions: improved artillery, portable and 'stronger than bronze'; and a device to prevent the sinking of ships 'even though breached by artillery in many places'. He had been interviewed by the Council of War two years before but 'I was not believed, they judging the things impossible; and not having won their confidence I returned home intending to think no more about it'. What happened next gives a further indication of the importance of piety in Philip II's Spain. In his letter to the king Nieba wrote: 'I told this to my confessor who knowing what I communicated to him to be the truth, instructed me to inform Your Majesty so that so useful a device should not be lost'. The result was that the inventions were referred for renewed examination by the captain-general of artillery and the secretary of war for the sea.[157] The proven integrity of the inventor had clearly counted here.

The historical record of the crown's processing of inventions presents few instances of clear adoption; even when this was encouraged, as in the successful testing of Juan Romeo's improved match for arquebuses,[158] the evidence of the king's authorisation for manufacture and supply is frequently lacking. But the evidence is complete for Antonio Marin's 'artifice, in the shape of a column and similar to an arquebus, which fires shot and darts with much impetus and effect, and without expense in powder or danger of fire, and many other advantages'. After its examination by the Council of War the king granted the inventor a monopoly of manufacture and sale of the gun in Castile for a period of fifteen years.[159] And a similar monopoly was conceded for a new technique of careening, the cleaning and repairing of ships' bottoms, indispensable for war and commerce. In 1560 the king had given a ten-year privilege to Giacomo de Francisco for his secret method which avoided the structural strains imposed by the common practice of laying ships on casks or boats. And when the privilege was about to expire the king asked the *Casa de la Contratación* for a report on its results. The reply indicated that the new technique was 'greatly appreciated' for its speed, economy and lack of damage to ships. By this time the inventor had left Spain, and because Philip was concerned to maintain the technique he granted an extension of the privilege to Pablo Matias and Andrea Barrasi, who had acquired the secret, appointing them jointly as chief careener with the right to charge up to 20 ducats for each 100 *toneladas* of shipping treated. They worked in Seville, the principal centre for careening, and were supplied by the *Casa de la Contratación* with the necessary equipment. The details of the new process are not revealed, but may have involved scorching.[160]

The fullest information on the crown's testing of military inventions concern projects which ended in failure; the documents reveal the considerable care with which tests were conducted. In one case the tests are recorded in great detail because the procedure was challenged in a lawsuit. The invention concerned a remedy for one of the greatest enemies of wooden ships – the worm, which had made a honeycomb of the ships taken in Columbus' fourth voyage, and which shortened the life and threatened the safety of every ship. Lead sheathing, known already to the ancient Greeks, was an expensive remedy and too heavy to use on swift galleys; but it was used on Spanish ships sailing to the Indies.[161] Tallow and pitch were also tried. The application of fire during careening was

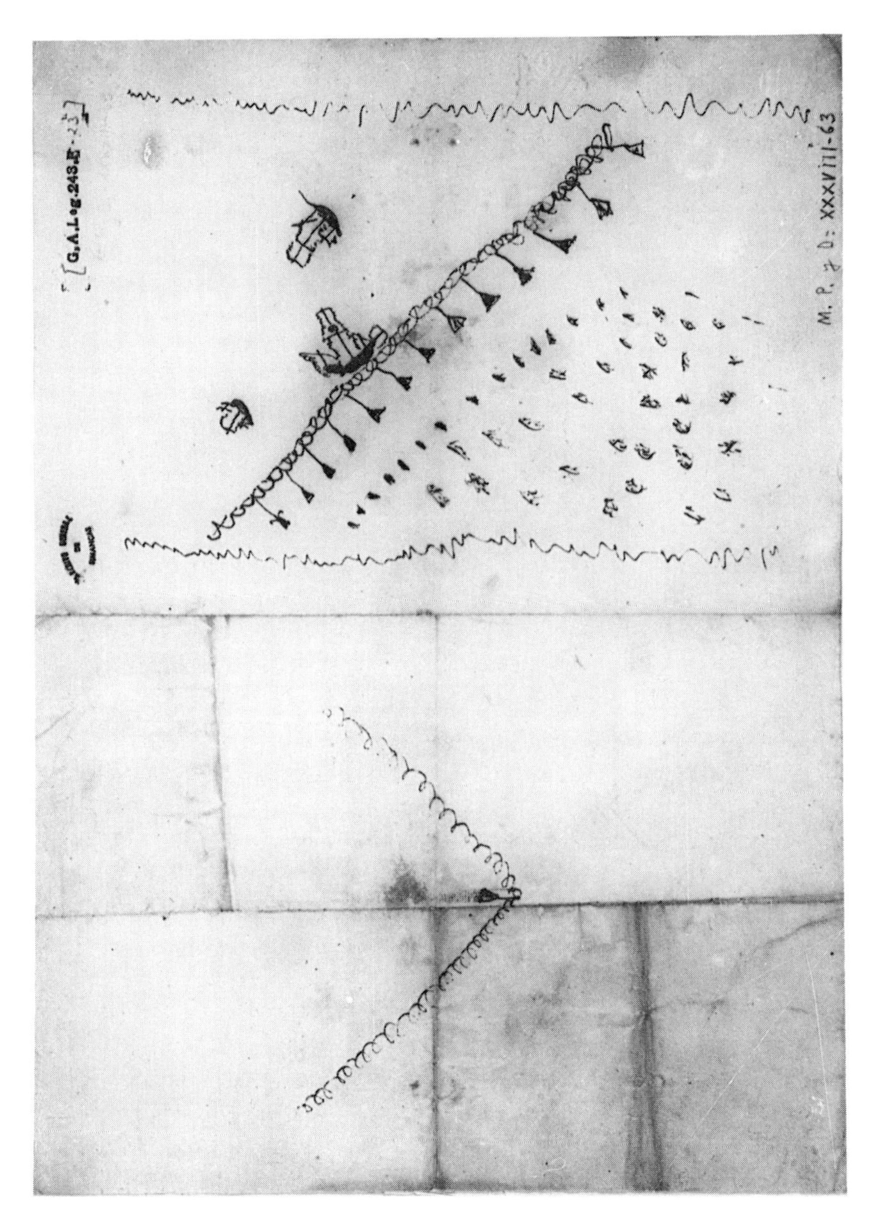


Plate 6 Sketch of a proposed chain barrier to protect armadas from fire-ships. The inventor, Clemente de Diomelodiede, a Neapolitan ensign, was given leave from his regiment to demonstrate his device to the king's ministers.
(AGS:GA 243/23; *Mapas, Planos y Dibujos* XXXVIII–63; 1588).

thought to kill any worm in the ship's planking and to prevent further boring. It was this technique which two Venetians, Nadalin Olivo and his associate Vicente Paletino, a Dominican, sought to replace with their discovery of an improved bitumen. They had first offered it to Charles V to protect warships and merchant shipping. It was agreed that an experiment should be performed at the Venetians' expense. When Philip succeeded his father he wrote to the officials of the *Casa de la Contratación* asking for a meeting of experts to be arranged to discuss the planning of the experiment.[162]

The experts, members of Seville's *Universidad de Mareantes*, the institution of shipowners, masters and pilots involved in the Indies traffic, decided on a controlled experiment.[163] 24 oak planks would be taken to Santo Domingo, Vera Cruz and Nombre de Dios, half of them scorched against the worm, the other half coated with the Venetians' new bitumen and carrying identifying marks. The worm would be given every opportunity of getting to the planks: some planks were nailed to the hulls of the ships sent from Seville; others were carried on board and on arrival in the Indies were submerged for long periods – nine months at Vera Cruz – in waters suspected to have abundant worm. Finally the experimental planks would be returned to Seville to see which had best survived the ravages of the worm.

From start to finish Olivo and Paletino protested that vested interests had prevented a fair trial of this experiment. They accused the *Casa* of negligence in assigning 12 of the planks to Pablo Matias, whom they alleged was an interested party because his occupation of careening would be threatened by the new technique. And what had he done to the planks? He had not burned them slightly in accordance with usual practice, but had burned them to charcoal. The Venetians alleged that this would favour the existing technique because worms bore only into wood: 'whoever heard of a ship of pure charcoal!' Duarte, factor of the *Casa*, accepted that the planks had been excessively burnt and appointed a caulker to treat them in the usual way.

The *Casa* took precautions to prevent any tampering with the specimens. Masters were to ensure that in the Indies ports the planks were submerged and later taken out of the water in the presence of local justices, the ship's notary, and other witnesses; and that on the return journey the planks were put 'safely below deck in a place where no one can reach them'. Notaries were to inspect and record the condition of the planks after submersion. Some of the planks were never seen again. The officials of the *Casa* said it was not known what had happened to them, but that the losses were not surprising 'in view of the great passage of time and tide'. But most of the planks did return and some seamen gave evidence under oath, declaring that they had cut open the planks and found that those treated with the Venetians' remedy had been badly eaten, while the rest were unaffected. Olivo again suspected a conspiracy and refused to accept the evidence. He wanted his own nominees to do the inspection, and alleged that some of the planks examined were not the originals because the identifying seals were not visible. The *Casa* responded that these may have been removed by the water, but granted a second inspection, which was more favourable to the Venetians' case.

The lengthy proceedings came to a close in April 1560 when the *Casa* gave its verdict: the new bitumen was far inferior to the existing technique of protection against the worm. The Venetians accused the *Casa* of neglect and others of deceit and appealed to the Council of the Indies to conduct a fairer inspection of the planks; the appeal was turned down.

The crown also became interested in inventions to secure supplies of drinking water for military forces on land or sea. The anguish of thirst came soonest in Africa. When García de Toledo began fortifying the Peñon de Vélez, after its capture from the Barbary pirates (1564), he 'found that no difficulty was greater than the shortage of water'. And later when an expedition to conquer Algiers was under consideration, the king was impressed by the invention of a water container specially designed for this goal, and gave instructions for the plan to be retained 'since it could be useful'. The proposal was for the construction of 500 earthenware vessels, covered with leather and provided with iron rings so that they could be chained together in pairs and slung over the pack-saddles of mules; the vessels would be filled at the nearest source and then carried back to camp (plate 7).[164] Greater expectations were raised by a different method of relieving thirst: the production of drinking water by distilling the salt water of the sea. The Spanish forces may well have been the first to investigate this for military purposes. It was tried in June 1560 at Djerba where a force of 6000 Spanish, Italians and Germans, with little water and in the heat of

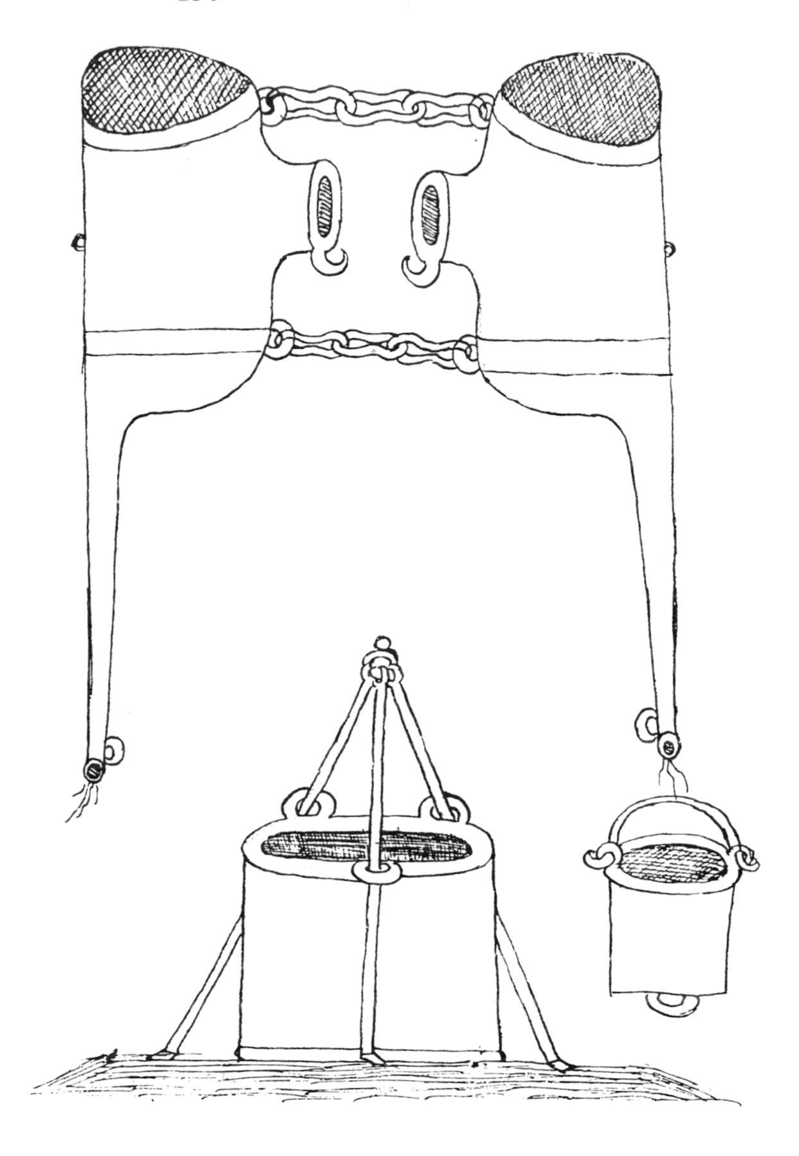

Plate 7 One of the numerous inventions which came before the Council of War. Chained pairs of water vessels, designed to facilitate the supply of drinking water to Spanish troops on a projected expedition to Algiers. (AGS:GA 77/166, 1573).

an African summer, was trapped by the Turks. But their commander Alvaro de Sande was hopeful because a Sicilian was using alembics to distil sea-water; and news soon reached the duke of Medinaceli, the viceroy of Sicily who had himself escaped from the beleaguered fortress, that 8–12 barrels of 'superior drinking water' were being produced by this technique every day. The duke informed the king that this was a 'miraculous discovery' which promised to be the salvation of the expeditionary force. The duke's confidence rested on a distillation experiment which he had arranged in his residence: the result was water as 'fresh as the Sierras of Medinaceli'. His concern that distillation at the fort would come to an end when fuel supplies ran out was lessened by the news that the wood of the galleys which had survived the naval disaster would be used for fuel. But in the end the duke's fears were realized. By early July Sande reported that soldiers were dying every day from lack of water. Rationing had been introduced and soldiers were digging for water. Wood for fuel was running out and there were not enough distillation vessels; of the 300 barrels of water required each day only 33 could be supplied by distillation. At the end of July there was no more water to drink and the fortress surrendered.[165]

It was not only Africa that presented Spanish military expeditions with urgent problems of water supply. Wherever ships sailed adequate provision was indispensable. The range of action of any galley was limited by the little space which could be given to the barrels of drinking water needed to satisfy the thirst of soldiers, seamen and the struggling oarsmen until a suitable watering place was within reach. It has been calculated that a typical galley of 144 oarsmen and 30 soldiers would be able to carry no more than 20 days' supply.[166] Nor did the water keep well; it putrefied after a few days into a disgusting turbid state. Wine lasted longer and was commonly used to supplement water, but it was unsuitable to give in large quantities to fighting men. When the Invincible was about to set sail Philip wrote to Medina Sidonia that since the wine of Andalucía was 'strong, and injurious to health, particularly for those who are not used to drinking it,' the usual wine ration should be reduced 'for the successful outcome of the enterprise'.[167]

Philip had recently issued instructions for 300 ducats to be paid to bring Luis Lorenza from Genoa to Seville to test his invention for making sea-water drinkable. Medina Sidonia was appointed to arrange this; it was June 1586 when he was busy with the preparation of the Invincible, though not yet its commander. The duke sent a municipal officer and a notary to Bonanza to witness the test aboard a ship at anchor on the Guadalquivir. A search was first made of the vessel, as the king had instructed, to ensure that there was no trick to employ fresh water already on board. The charcoal fuel to be used was stacked to see how much of the ship's deck it took up. Then Lorenza lit his device, a boiler 'made of copper and iron, the size of a barrel, and enclosing a flame', and distillation of sea-water began. The experiment was quantitative: the weights of charcoal before and after the distillation were measured to see how much fuel had been consumed in a given time (3 *arrobas* in 4 hours); the volume of distillate was also measured (7 *arrobas*). Don Miguel de Eraso, general of the Tierra Firme fleet was also present and like the others drank the water and found it to be 'of good taste and without odour'. But although the witnesses agreed that the enclosed flame of the apparatus was safer than the open fire used to cook meals on board ship, Medina Sidonia decided that the danger of fire was still too great because 'nothing is safe where there is timber and pitch'. The duke also thought the apparatus and fuel would be an encumbrance; but he strongly recommended the invention for the use of forces on land. The duke's report went to the Council of the Indies where it was accepted and recommended to the king.[168]

Lorenza's apparatus was not used for the Invincible. That armada carried 11 000 barrels of water, intended to last for six months. But the barrels were of poor quality and leaked, the result of the loss of seasoned staves and hoops during Drake's attack at Sagres (1587). Adverse winds had slowed the armada's passage and just twenty days after setting sail Medina Sidonia was forced to enter La Coruña because of shortage of water, which he said was 'the thing which matters most'. The king put his trust in God 'that these initial difficulties will end in great glory'. But after the battle in the Channel, Medina Sidonia told the king of the suffering from food and water rationing. Some of the water casks contained green slime, and on the terrible journey back to Spain the duke ordered horses and mules to be thrown overboard to save drinking water; and for twelve days one crew lived off rainwater squeezed from their shirts.

The installation of an efficient and compact distillation plant on ships presented a complicated technological problem which was not solved until the nineteenth century. Sixteenth-century technology was inadequate for this as well as for the production of powerful, safe and accurate guns. Because of the limitations of existing technology, inventors had little influence on the wars of sixteenth-century Europe. Neither Spain nor her enemies held a decisive technological advantage. Many more of the Invincible's ships were destroyed by the weather than by English naval superiority. And if the Dutch had possessed any advantage in military technology the struggle against Spain in the Netherlands would not have lasted eighty years. As for Philip's ambition to make Spain independent of Europe for its military equipment there were too many obstacles in the way. Philip may have been the mightiest monarch on earth but he was not the master of peninsular Spain's mountainous terrain and unnavigable rivers, not the absolute ruler of its peoples, not in command of an abundance of native technological skill, and not in control of the economy.

## Notes

1. G. Parker, *The Army of Flanders and the Spanish Road 1567–1659* (Cambridge, 1972) pp. 293–5; H. Kamen, *Spain 1469–1714* (London and NY, 1983), p. 161f.
2. I. A. Thompson, *War and Government in Habsburg Spain 1560–1620* (London, 1976).
3. W. H. McNeill, *The Pursuit of Power* (Oxford, 1983), pp. 111–14.
4. C. H. Haring, *Trade and Navigation between Spain and the Indies* (Cambridge, Mass., 1918), p. 269. Similarly, a distinguished Spanish naval historian saw Philip as 'unsympathetic' to marine affairs and attributed this to the king's sea-sickness: C. Fernández Duro, *Armada española desde la unión de los reinos de Castilla y de Aragón*, vol. 3 (Madrid, 1897), p.173.
5. *Recopilación de las leyes destos reynos* (Alcalá, 1598), *lib.* vii, *título* vii, *leyes* vii and xv.
6. *Ibid.*, *lib.* vii, *título* vii, *ley* xvii. Independent local action to control cutting had already been taken in 1526 by the authorities in Vizcaya with a view to maintaining supplies for its important industries: timber for shipbuilding and charcoal for ironworks, T. Guiard y Larrauri, *La Industria naval vizcaina* (Bilbao, 1917), p. 58. And in 1547 local interests led the *Cortes* of Catalonia to prohibit the export of timber from the Principality in order to preserve the economic benefits of constructing galleys at Barcelona for the monarch; J. Carrera Pujal, *Historia política y económica de Catalunya*, vol. 2 (Barcelona, 1947), p. 362.
7. AGS:GA 347/19, royal *cédula*, 6 May 1563.
8. MN:SB 3/184, royal *cédula*, 7 December 1564.
9. MN:FN XXII/28/86, Cristóbal de Barros to the king (n.d.).

10. *Ibid.*, f.87.
11. The superintendency is first mentioned in the *cédula* of 7 December 1564 (note 8); the full extent of Barros' powers is indicated in AGS:GA 347/57, royal *cédula*, 7 December 1574, Madrid.
12. AGS:GA 82/223, the king to the *regente* and *alcaldes mayores* of Galicia, 5 August 1577. Local justices even obstructed the king's orders to cut trees in royal forests. The building of the mole at Málaga to shelter galleys from storms was delayed by the refusal of the *alcalde mayor* of Marbella to allow timber to be cut without payment; the municipality had been profiting from sales of the timber even though it was royal property: *ibid.*, 287/231 and 287/236–9, correspondence of July–August 1590.
13. *Ibid.*, 72/298 (n.d.).
14. *Ibid.*, 262/202, *consulta* of the *Consejo de Guerra*, 10 May 1589.
15. *Ibid.*, 82/229, draft of a letter from the king to Barros, 1577.
16. J. Carrera Pujal, *Historia de la economía española*, vol. 2 (Barcelona, 1944), p. 218.
17. J. Reglá Campistol, *Felip II i Catalunya* (Barcelona, 1956), p. 43.
18. AGS:GA 81/315 and 208/251, *consultas* of the *Consejo de Guerra*, 1576 and 23 April 1587; and *ibid.*, 322/157, Maestre de Montesa to the king, 16 June 1591, Barcelona.
19. *Ibid.*, 264/219 and 264/227, reports by Antonio de Alzatte on galley construction at Barcelona, 1589. It cost nine *reales* to transport a piece of timber over these four leagues by land, and four *reales* for the nine leagues by sea. A route entirely by land, though more expensive, was sometimes preferred for reasons of safety for transporting timber for galleys of special importance; *ibid.*, 195/136, report by Pedro de Isunza, *veedor* and *contador*, 20 December 1586, Barcelona. This report also advised the king against exploiting the rich pine forests of Rosellón because, while the land there was flat, the population, 'French' and potentially hostile, would impede cutting; also the snows would hinder cutting and transport for several months of the year.
20. *Ibid.*, 186/73, Pedro de Isunza to the king, 6 May 1586, Barcelona.
21. *Ibid.*, 283/304 same to the same, 6 April 1590, San Felieu de Guijoles; and *ibid.*, 322/157 Maestre de Montesa to the king, 16 June 1591, Barcelona. Exactly the same cause of disruption of supplies created shortages of timber for gun carriages. Cut timber, estimated to be worth 30 000 ducats, was rotting for up to six years in Murcia and Andalucía, because of failure to supply payment for transport to munitions houses at Málaga and Cartagena: *ibid.*, 151/354 Francés de Alava to unnamed person, 9 November 1583.
22. *Ibid.*, 249/241, Antonio de Alzatte to a royal secretary, 22 June 1589, Madrid. The construction of the road is mentioned in *ibid.*, 264/ 234, Felipe de Queralte, governor of Castilbo, to the king (n.d.). The complaints concerning the failure to transport masts were expressed in the above letter from Alzatte. A plan to float pine logs from the Pyrenees along the rivers Zinca and Ebro to supply the arsenal at Barcelona at much reduced transport costs was suggested by Jaime Tanegas, an architect of Zaragoza, in a letter to the governor of that city: MN:FN XII/74/265–7 (n.d.).
23. AGS:GA 77/11, the king to the *regente* of Galicia, 17 May 1572.
24. *Ibid.*, 245/165, Manrique de Lara to the king, 27 February 1589, Barcelona.
25. *Ibid.*, 246/281, report by Alzatte, 24 March 1589, Barcelona. The report also found Vinaroz, the intended site for construction, too exposed to corsairs and too poorly equipped with naval stores. A startling statement in the report indicated that experts from Barcelona had already surveyed the same forests in 1562 and found the timber unsuitable. Why didn't someone make this known before the second expedition was sent?
26. F. Olesa Muñido, *La organización naval de los estados mediterraneos y en especial de España durante los siglos XVI y XVII* (Madrid, 1968), p. 631.
27. Shortage of masts for twelve new galleons built at Bilbao and Guarnizo led to a contract with a German to import them 'from Danzig or Germany' at a cost

of 5000–6000 ducats; AGS:GA 285/168 Bernabé de Pedroso to the king, 9 June 1590, El Ferrol.

28. AGS:GA 72/300, royal *cédula*, 7 August 1568.
29. *Codoin*, 92, pp. 151–2; Bernardino de Mendoza to the king, 20 October 1581, London.
30. The naval recovery of the 1560s is discussed by F. Braudel, *The Mediterranean and the Mediterranean World in the Age of Philip II*, trans. S. Reynolds, vol. 2 (London, 1973). p. 1007f; the part played by two viceroys, García de Toledo and Hurtado de Mendoza, is illustrated in Reglá, *Felip II i Catalunya*, p. 43f.
31. MN:SB 3/666, the king to duque de Maqueda, viceroy of Catalonia, 27 May 1594, Madrid. At the same time similar instructions were sent to the viceroy of Naples to engage the best craftsmen and begin construction at the royal arsenal there; the king had received reports that the Turkish fleet was about to go on the offensive; *ibid.*, 3/660, the king to conde de Miranda, 4 May 1594, Madrid.
32. AGS:GA 189/96, Pedro de Isunza to the king, 25 December 1586, Barcelona.
33. *Ibid.*, 202/27; 204/212–13; and 225/223, Manrique de Lara to the king, 8 October 1587; 23 December 1587; and 10 July 1588, Barcelona.
34. MN:SB 4/937, Pedro de Isunza, 'Relación del dinero que ha venido de Castilla,' 6 September 1588, Barcelona.
35. AGS:GA 202/27 and 245/165 Manrique de Lara to the king, 8 October 1587 and 27 February 1589, Barcelona; MN:SB 3/567, Alzatte to the king (n.d.). Apart from the cost of construction the king had to meet annual costs of a similar order for galley maintenance: careening, repairing planks, renewing oakum; and to feed and pay the large number of seamen and soldiers which each galley carried. In one year of the reign the cost of maintenance amounted to 5653 ducats for each galley: MN:FN XII/84/313–8, 'Relación del gasto que una galera hace en un año,'(n.d.).
36. AGS:GA 165/276, 'Consulta de los cargos y condenaciones que fueron hechos y se hazen a algunos de los oficiales que por su mag [estad] entendieron en el beneficio de la fabrica de galeras,' (1584). There had been earlier attempts to introduce rigorous records of carters' loads, paying according to actual deliveries instead of what had been promised, for which 'much care is needed because it often happens that only half a cart-load is delivered'; MN:FN XII/73/434, 'Relación de lo que parece se ha de hacer en Barcelona por el señor Mosen Francés Setanti a cuyo cargo esta la fabrica de las galeras,' 20 January 1555.
37. AGS:GA 264/243, from registers kept by Pedro de Isunza, *veedor* and *contador* of Catalonia, 23 April 1589.
38. *Ibid.*, 264/245, from the registers of Pedro de Isunza, 26 March 1589.
39. *Ibid.*, 226/173, Pedro de Isunza to the king, 9 August 1588, Barcelona.
40. *Ibid.*, 179/299, Pedro de Isunza to unnamed official, 5 December 1585; *ibid.*, 179/309, Juan de Carlua to the king, 23 December 1585, Barcelona; MN:SB 4/1225, 'Relación dellos maestros mayores. . .que han servido en las atarazanas de esta ciudad de Barcelona,' 30 May 1595, Barcelona.
41. AGS:GA 264/219, (1589); and related accounts in *ibid.*, 264/220 and 225.
42. *Ibid.*, 285/355, the *veguer* of Barcelona to the king, 29 June 1590, Barcelona. Because of the pestilence it was suggested that galley construction be continued at Blanes: *ibid.*, 264/231, various recommendations by Manrique de Lara and Alzatte, 1589.
43. *Ibid.*, 264/223 for Alzatte's estimate, 29 July 1589; MN:SB 4/1162 for the account of construction in 1591–3 given by Antonio de Irabien and Sebastián de Ruy García, 10 March 1594. My suggestion of a temporary reduction in costs does not agree with Thompson's conclusion that costs were steadily rising in 1583–1601: *War and Government in Habsburg Spain*, p. 331, fn. 46.
44. *Codoin*, 29, p. 35, García de Toledo to duque de Francavila, viceroy of Catalonia, 31 January 1565, Baya; *ibid.*, 29, p. 555, García de Toledo to the king, 19 October 1565, Messina. In this letter to the king he complained that in Barcelona galleys

used to be made 'too bulky' and 'now in trying to remedy this they have made them too small.'

45. AGS:GA 202/27, Manrique de Lara to the king, 8 October 1587, Barcelona.
46. MN:SB 5/61, 'Relación de lo que esta concertado con maestre Geronimo Verde,' November 1597 ; and *ibid.*, 4/1207, a related letter from Gian Andrea Doria, 12 December 1594, Genoa.
47. AGS:GA 264/244, Isunza to the king, 7 May 1589, Barcelona.
48. MN:SB 4/464, Carlo Brancazzo to the king, 6 January 1575, Madrid.
49. AGS:E 1077/37, marqués de Mondéjar to the king, 24 February 1578, Naples.
50. Reglá *Felip II i Catalunya*, p. 47, referring to an order of the viceroy, Hurtado de Mendoza, on 21 November 1566.
51. AGS:GA 70/296, the king to Capitán Martín de Arana, 18 June 1562, Madrid.
52. For an example of this, MN:SB 3/442, the king to Cristóbal de Barros, 1581, Lisbon; this related to the construction of galleons on the north coast.
53. Reglá, *op.cit.*, p. 46 referring to Hurtado de Mendoza's order of 26 August 1566.
54. AGS:GA 219/158, Manrique de Lara to the king, 13 January 1588, Barcelona.
55. *Ibid.*, 264/241, Don Pedro de Cardona to the king, 26 May 1589, Barcelona.
56. Carrera Pujal, *Historia política y económica de Catalunya*, vol. 2 (Barcelona, 1947), p. 362.
57. AGS:GA 264/235 (n.d.).
58. *Ibid.*, 248/51, Agustín de Hojeda to the king, 19 May 1589, Deusto.
59. *Ibid.*, 195/14 'Relación del estado que estan la madera y demas materiales que ay en las Atarazanas de Barcelona,' 20 January 1586.
60. From a document in Carrera Pujal, *op.cit.*, vol. 2, p. 362. The document states that before its introduction much timber had been 'lost', which leaves the function of the process vague.
61. MN:SB 5/26, *asiento* with Álvaro de Bazán, 1540.
62. J. de Escalante de Mendoza, 'Itinerario de Navegación,' in C. Fernández Duro, *Disquisiciones náuticas*, vol. 5 (Madrid, 1880), p. 413. The MS. of around 1575 was addressed to the king.
63. MN:FN XXII/76/293, document of 19 March 1581, recording a meeting of experts in Santander including Pedro de Busturia senior and junior, the two men who built Menéndez's galleons; they stated that the dimensions of these ships were: length 44 cubits, breadth 12–13 cubits.
64. MN:SB 3/649, the king to Cristóbal de Barros, 6 May 1563, Madrid.
65. MN:VP 49/229–30, royal *cédula*, 26 July 1582, Lisbon, extending the concession for a further ten years.
66. AGS:GA 72/300, royal *cédula*, 7 August 1568; the Catholic Monarchs had given subsidies for ships of 600 *toneladas* upwards. The '*tonelada*', equivalent to about nine tenths of the English 'ton,' was an Andalusian measure of cargo capacity equal to two standard wine casks ('*pipas*') of Seville. The *tonelada* was one fifth smaller than the *tonel*, the unit employed in Vizcaya.
67. MN:FN XXII/28/90, Barros to the king (n.d.).
68. *Ibid.*, *ibid.*, f.95v–100.
69. The correspondence between Barros and the king, and the views of 'experts' on this matter are recorded in MN:FN XXII/76/286–329.
70. *Ibid.*, *ibid.*, f. 328. It seems however that the extensive cuts brought complaints because when other galleons were later constructed on the coast the king ordered the work to be done on three separate sites since this 'reduces the damage to the locality, avoiding the consumption of so much wood that would occur if all was done on a single site;' MN:SB 3/556, the king to Juan de Cardona, 29 January 1589, Madrid.
71. AGS:GA 142/185, Barros to a royal secretary, 27 February 1583, Guarnizo.
72. MN:FN XXII/28/108–108v, Barros to the king, (n.d.). The fields of Calatayud and Calahorra were capable of supplying hemp in considerable quantities; in 1590 because of abundant rainfall a bumper harvest was expected to provide 2000 *quintales* for rigging six new galleons under construction at Deusto; AGS:GA

284/22, Agustín de Hojeda to the king, 25 May 1590. Later that year hemp from the same area was keeping ropemakers busy making cordage for twelve other galleons at Portugalete (Bilbao); the hemp was said to be expensive but a good investment because cables made from it lasted 'five or six years while a cable of Flanders hemp was useless after one year; and shipowners on this coast only trust hemp from Calatayud'; AGS:GA 290/161, Baltasar de la Cama to the king, 20 November 1590, Portugalete. According to Thompson the local hemp industry collapsed after 1590 when the ropemakers of Bilbao, working for the crown, were ruined after they were left unpaid for the hemp and pitch which they had purchased; *War and Government in Habsburg Spain*, p. 213. Hemp was also in demand for arquebus match; in 1590 2500 *arrobas* were collected within the Peninsula for that purpose; AGS:GA 280/256, Pedro de Izaguirre to the king, 20 January 1590, Pamplona.

73. AGS:GA 142/185, Barros to a royal secretary, 27 February 1583, Guarnizo.
74. MN:SB 4/1184 and 5/55; concerning the proposal and implementation of Pedro López de Soto's design of five ships totalling 600 *toneladas* for the defence of the Portuguese coast, August 1594 – January 1595.
75. AGS:GA 285/162, Alonso de Bazán to the king, 23 June 1590, El Ferrol.
76. MN:SB 3/556, the king to Juan de Cardona, 29 January 1589, Madrid; AGS:GA 281/52, Juan de Soto to the king, 12 February 1590, Guarnizo.
77. AGS:GA 199/47, Juan de Acuña Vela to the king, 15 July 1587, Lisbon.
78. *Ibid.*, 80/192, *consulta, Consejo de Guerra*, 1575.
79. *Ibid.*, 222/51, Juan de Acuña Vela to the king, 5 March, 1588, Lisbon.
80. Luis Collado, *Platica manual de artillería*, (Milan, 1592), p.8.
81. AGS:GA 220/14 Juan de Acuña Vela to the king, 20 February 1588, Lisbon.
82. *Codoin,* 111, pp. 482–3, the king to Francisco Hurtado de Mendoza, ambassador in Vienna, 23 October 1574, Madrid.
83. AGS:GA 88/151, Francés de Alava, captain-general of artillery, to Juan Delgado, secretary of war, 4 March 1578 on the abortive attempt of Diego Sobrino, a Portuguese, to make cast-iron cannon.
84. AGI:GIG 743/64, *consultas* of the *Consejo de las Indias* and correspondence relating to Rodi's proposal, November 1592-June 1595.
85. AGS:GA 78/321, statement by Francés de Alava, 17 March, 1574.
86. *Ibid.*, 88/324, Francés de Alava to the king, 4 November 1578, Madrid.
87. Real Academia de Historia MS., Colección Muñoz, A/111 f.215 and A/112 ff.331v–2; letters to the emperor from officials in Santiago de Cuba, September 1547 and July 1550. Prior to this contract Cuban copper had been shipped to Spain in 1541 and was used to make two pieces of bronze artillery; AGI:*sección de Contaduría, legajo* 275, *documento* 1, fol. 141v, c.1550.
88. AGI:P 238ii/1, contract between the king and Sancho de Medina, 20 January 1578, Madrid.
89. AGS:GA 88/328, *consulta, Consejo de Guerra*, 11 November 1578 with the king's annotated response.
90. AGI:GIG 744/55, *consulta, Consejo de las Indias*, 8 September 1596, Madrid. The Council also supported schemes to work copper mines in Cartagena (de Indias) but rejected the governor's suggestion that a labour force of Moorish galley slaves be used, because 'they might infect the negroes and Indians with the evil customs of their sect,' *ibid.*, 745/86, *consulta, Consejo de las Indias*, 22 June 1598, Madrid.
91. AGI:*Gobierno: Santo Domingo, legajo* 451, *documentos* 6–7c, correspondence of 1607 relating to the deficiencies of Cuban foundries.
92. AGS:E 1324/49, Juan de Spinosa to duque de Sessa, governor of Milan, 26 December 1560, Venice. An alchemical treatise stated 'that one metal can be converted into another is clearly demonstrated by Paracelsus. For he says the inhabitants of one locality in Germany put thin plates of iron into a mineral water of copperas [copper sulphate] and find after some months that it has been converted into copper,' BNM MS.2058: Richard Stanyhurst, 'Toque de Alchimia,' (1593), f.252. The supposed interconversion of elements was therefore

merely a precipitation of copper from a solution of one of its salts by the addition of iron.

93. T. González, *Registro y relación general de minas de la Corona de Castilla*, vol. 2 (Madrid, 1832), pp. 370–8.
94. *Ibid.*, vol. 2, pp. 517–8.
95. AGI:P 251/71, evidence given to officials of the *Casa de la Contratación*, 7 June 1573, Seville. Juan Morel worked in a small foundry established in Seville in 1565 to supply the Indies fleet with artillery. Fear of disputes over claims had also deterred landowners from operating mines (of unspecified ore) in Valencia; Carrera Pujal, *Historia de la economía española*, vol. 2, p.510.
96. AGS:GA 190/490; 208/382; 215/16; 215/17; and AGS:DC 8/57 concern Cipriano's project.
97. Tin ore was much less abundant in Spain than copper. There were occasional reports of discoveries of tin mines with subsequent signs of royal interest; AGS:DC 8/53, the king to Juan Osorio de Valdes, draft letter 10 July 1580, Madrid. But England was the regular source of supply of tin and lead even in the 1580s when, despite the growing hostilities, tin imports continued and England received iron from Vizcaya: M. Ulloa, 'Unas notas sobre el comercio y la navegación españoles en el siglo XVI,' *Anuario de Historia Económica y Social*, **2** (1969), 191–237. Even in the worsening relations with England during 1587 four bronze culverins for the defence of the Canaries were made in Seville from English tin; AGS: *Mapas, Planos y Dibujos*, V-16, instructions of Acuña Vela, 25 July 1587, Lisbon. The English Levant Company also supplied the Turks with tin and copper; when in 1580 the viceroy of Sicily was informed that English ships carrying this merchandise were calling at Sicily en route to Constantinople, he told Philip he was inclined not to seize the cargo to avoid giving Elizabeth 'cause for complaint'; AGS:E 1149/39, Marc Antonio Colonna to the king, 26 February 1580, Palermo. But when the Invincible was being prepared, and there was no tin for the Lisbon foundry, the captain-general of artillery had no hesitation in seizing a cargo of tin and lead when a Venetian ship returning from England called at Belém; AGS:GA 220/12 Juan de Acuña Vela to the king, 13 February 1588, Lisbon. Tin for the Invincible's artillery was also secured by *asientos* with French merchants.
   AGS:GA 220/12, Juan de Acuña Vela to the king, 13 February 1588, Lisbon. Tin of smelting also tested Italian founders at Málaga; their experiments were disrupted by the cracking of a furnace; *ibid.*, 284/265, same to the same, 9 May 1590, Málaga.
99. González, *op.cit.*, vol. 2, pp. 23–9 for Schedler's appointment and his letter to the king of 1595.
100. *Ibid.*, vol. 2, pp. 49–54 for the king's instructions to Ayanz, 8 July 1597.
101. C. Clay, *Economic Expansion and Social Change: England 1500–1700*, vol. 2 (Cambridge, 1984), p. 60.
102. AGS:GA 67/201, document of 1558 relating to Vidal's arrest; *ibid.*, 82/18, Francés de Alava to the king, 17 July 1577, Madrid, on Martin's activities. There were other allegations of fraud in saltpetre or gunpowder manufacture for the crown at Tembleque, and also in Sicily and Milan: *ibid.*, 77/204, the king to Agustín de Zárate, 26 October 1572; AGS:E 1243/12, Marc Antonio Arroyo to the king, 7 April 1575, Milan; *ibid.*, 1144/75, report of Bernardino de Velasco, 1575.
103. AGS:GA 189/92; the king to the marqués de Almazán, 2 October 1586, Escorial; with reply of Luis Carillo y Toledo, 16 December 1586, Pamplona.
104. *Ibid.*, 170/160 describes the technique to be used in refining saltpetre, 1584; *ibid.*, 222/117, Pedro de Pinedo to the king, 24 March 1588, Málaga, expressing dissatisfaction with the saltpetre received.
105. *Ibid.*, 156/56, draft of royal *cédula*, 1583.
106. *Ibid.*, 323/144, Pedro de Bracamonte to the king, 5 August 1591, Cartagena.
107. *Ibid.*, 200/57, Pedro de Pinedo to the king, 1 August 1587, Málaga.
108. *Ibid.*, 76/101, drafts of letters from the king, 8 July 1572.

109. *Ibid.*, 209/367; 211/3; 234/65; 234/242, *consultas, Consejo de Guerra*, 27 April 1587; 26 October 1587; 5 February 1588; 30 May 1588 concerning the inspection and purchase of the sulphur mines. *Ibid.*, 266/63 records the agreement between the king and Monreal, 6 May 1589, Escorial.
110. *Ibid.*, 254/219, Acuña Vela to the king, 14 December 1589, Cartagena.
111. *Ibid.*, 317/219, Monreal and Alonso Carrasco to the king, 1591.
112. *Ibid.*, 317/218, Andrés de Angulo, governor of Caravaca, to the king, 16 January 1591.
113. *Ibid.*, 317/220, Monreal and Carrasco to the king, 23 November 1590, Hellín.
114. *Ibid.*, 290/264, Carrasco to the king, 24 November 1590, Hellín, with Prada's annotation.
115. *Ibid.*, 317/222, Carrasco to the king, 18 January 1591, Hellín.
116. González, *op.cit.*, vol. 2, p. 53. In 1593 Hellín was able to produce 50 *quintales* of sulphur for the Málaga powder-mills, and at about the same time 300 *quintales* for gunpowder manufacture at Pamplona; AGS:*Contaduría Mayor de Cuentas* (2a *época*), *legajo* 425, unfoliated. The same *legajo* contains records of accounts for the mine in 1593–1601; they indicate high production costs.
117. González, *op.cit.*, vol. 2, pp. 392, 394–5, 405, 410, 417. These documents show that in 1627 funds were granted to enlarge the plant to allow production of 2000 *quintales* a year; after nine months about 1250 *quintales* had been produced.
118. AGS:GA 186/118, Pedro de Pinedo to the king, 1 July 1586, Málaga.
119. *Ibid.*, 198/84, residents of Málaga to the king, 23 June 1587. Production at Málaga had fallen below the level reached 30 years before when 'a new engine' produced ten *quintales* a day; four men were then employed, operating 'a wheel and hammers which fall on mortars' so grinding the powder; *ibid.*, 68/25, the *corregidor* of Málaga to Juana, regent of Castile, 20 January 1558.
120. J. Vigón, *Historia de la artillería española*, vol. 1 (Madrid, 1947), p. 326.
121. B M Egerton MS. 1506 f. 217, the king to the *Suprema*, 16 November 1594.
122. BM Add. MS. 28, 353 ff. 146–62, Fernando de Torres y Portugal, *asistente* of Seville, to the king, 20 May 1579. The officials of the *Casa de la Contratación* were later blaming a local Morisco for purposely making ineffective gunpowder for the armada: AGS:GA 280/216, Juan de Ibarra to the king, 17 January 1590, Seville.
123. V. Vazquez de Prada, 'La industria siderurgica en España (1500–1650)', *Schwerpunkte der Eisen gewinnung und Eisen verarbeitung in Europa 1500–1650*, ed. H. Kellenbenz (Cologne, 1974), pp. 35–78.
124. AGS:E 1234/138, the king to the *comendador mayor* of Castile, 1 July 1572, el Pardo.
125. *Ibid.*, *ibid.*, 1239/220 and 221, letters of the king to the marqués de Ayamonte and the duke of Florence, 20 October and 20 December 1574.
126. AGS:GA 186/142, *consulta*, 20 July 1586; *ibid.*, 209/106, *consulta, Consejo de Guerra*, 1 August 1587.
127. *Ibid.*, 246/83 and 247/225, Lope de Echariz to the king, 14 March and 30 April 1589, Pamplona; Thompson, *War and Government in Habsburg Spain*, p. 253 f. discusses the economic failure of Euguí.
128. AGS:GA 269/11, Gaspar de Morales to the king, 19 May 1589; *ibid.*, 220/15, Acuña Vela to the king, 25 February 1588, Lisbon.
129. *Ibid.*, 82/174, Alava to the king, 17 October 1577, Madrid.
130. *Ibid.*, 198/32, Acuña Vela to the king, 30 May 1587, Lisbon.
131. AGI: *Contratación, legajo* 4871, *número* 1, records the successful examination of Juan Gutíerrez, native of Antwerp and resident of Mexico, on 18 May 1581; regulations of the school and specimen examination questions are given in a document reproduced in C. Fernández Duro, *Disquisiciones náuticas*, vol. 6 (Madrid, 1880), p. 438f. At the same time Philip was trying to refound the artillery school in Milan to supply gunners for Italy, Sicily and Sardinia. Again the intention was to secure exclusive employment of the king's subjects – 14 German gunners who had served the king in Italy had refused to go to Sardinia and could

not be compelled to obey. The suggestion that the school would attract trainees if they were granted the privilege of bearing arms caused some concern that ruffians would enrol. Local officials in Milan obstructed the plans which continued to be discussed in Philip III's reign. AGS:E 1245/62; 1245/64; 1249/80 and 81; 1260/117; and 1293/88 document the proposal and correspondence relating to it.

132. V. de la Fuente, *Historia de las universidades, colegios y demas establecimientos de enseñanza en España*, vol. 2 (Madrid, 1885), p. 485.

133. The Spanish treatises of the early-seventeenth century contained some of the first tables of ranges; and Diego Ufano's treatise of 1613 contains, according to A.R. Hall, 'the first attempt to define the trajectory of a projectile mathematically,' *Ballistics in the Seventeenth Century*, (Cambridge, 1952), pp. 45–7.

134. AGS:GA 281/230, Acuña Vela to the king, 12 February 1590, Málaga. Perhaps something of his undue confidence in mathematics is also evident in the precise arithmetical rules which he prescribed for calculating the quantity of metal required to found a piece of artillery – multiplying the diameter of the bore by four – and for estimating the charge of powder needed for guns of a particular calibre; BNM MS. 9027: G. Firuffino, 'Tratado de artillería,' (1599).

135. AGS:GA 280/228, duque de Medina Sidonia to the king, 16 January 1590, San Lucar; *ibid.*, 291/196–197, conde de Santa Agueda to the king, 10 December 1590, Puerto de Santa Maria, with the petition from the German gunners.

136. *Ibid.*, 80/84, G. B. Antonelli to the king, 19 December 1575, Madrid. Some salaries of engineers, much higher than those of physicians, university professors or cartographers, are given in J. M. López Piñero, *Ciencia y técnica en la sociedad española de los siglos XVI y XVII* (Barcelona, 1979), pp. 83–7, indicating a range of 300 000–700 000 *maravedís* a year for royal engineers; but his figures for ordinary gunners (20 000–50 000 *maravedís*) are higher than the sources warrant.

137. AGS:E 1212/180 the king to the duke of Savoy, 13 October 1561, Madrid.

138. *Ibid.*, *ibid.*, 1064/18–19, Granvelle to the king, 3 April 1574, Naples.

139. *Ibid.*, *ibid.*, 1324/113, García Hernández to the king, 11 June 1563, Venice.

140. *Ibid.*, *ibid.*, 1270/168, Pedro de Padilla to the king, 4 September 1592, Milan. The king had been making inquiries about him since 1574; AGS:GA 81/101, marqués de Ayamonte to the king, 15 March 1576, Milan.

141. *Ibid.*, *ibid.*, 1230/64, Juan de Vargas Mexia to the king, 23 April 1572, with the king's annotation.

142. AGS:GA 111/218, Francés de Alava to the king, 9 March 1581, Lisbon.

143. Further details of the Italian engineers are given in D. Goodman, 'Philip II's Patronage of Science and Engineering', *British Journal for the History of Science*, **16** (1983), 58f; and their activities in the Netherlands are discussed by C. Duffy, *Siege Warfare* (London, 1979), c.4.

144. Goodman, *op.cit.*, pp. 56–7.

145. Cristóbal de Rojas, *Teoría y práctica de fortificación* (Madrid, 1598), prologue.

146. J. Hale, 'The Argument of some Military Title Pages of the Renaissance,' *Newberry Library Bulletin*, **6** (1964), 92; and Hale, *Renaissance Fortification: Art or Engineering?* (London, 1977), p. 33f. for opinion in contemporary Italy on the relative importance of mathematics and military experience.

146a. Richard Butiler (Butler?) had come to Spain with two Irish companions to support the Catholic cause, but had been arrested on suspicion of espionage. When he learned from a pilot, who visited him in prison, that the king wanted to construct docks similar to those in England at Lisbon and other ports, Butiler claimed that he had designed precisely this device and immediately offered it in exchange for his freedom. But while his plan was approved it was the pilot who got the reward, leaving Butiler in prison protesting and alleging that he had not yet revealed the essential part of the secret design; AGS:E 170/2, Captain Richard Butiler to the king, royal prison of Madrid, 9 October 1594.

147. IVDJ 25J/223, Benito López de Gamboa to Mateo Vázquez, 10 April 1577.

148. AGS:E 1244/161, Giovanni Bianchi to the king, 15 January 1576, Pavia.

149. *Ibid.*, *ibid.*, 1333/71 and 72, Guzman de Silva to the king endorsing Dragovic's invention, Venice, 1574; *ibid.*, *ibid.*, 1508/27, Luis de Requesens to Guzman de Silva, 8 November 1574, Brussels.
150. AGS:GA 281/235, Acuña Vela to the king, 25 February 1590. Málaga.
151. *Ibid.*, 254/123, Martín Rico to the king, 23 December 1589, El Ferrol. In 1539–40 Charles V had given 100 000 *maravedís* to the Spanish mariner Blasco de Garay for a means of propelling ships by wheels; the device is said to have driven a boat three or four leagues in two hours, López Piñero, *Ciencia y técnica*, pp. 84 and 210–11.
152. AGS:GA 286/199, López de Chaves, *corregidor* of Málaga to the king, 9 July 1590, Málaga. In this the invention was described as 'laughable' because 'its poor design would hinder a ship's movement by the wind.'
153. IVDJ 95/136, Gonzalo Venegas to the king (n.d.). In this document Venegas described himself as a doctor in canon law and *referendario* to the ecclesiastical court in Rome.
154. AGS:E 1264/117 and 118, Tirone's proposal and the result of its consideration by the *Consejo de Guerra*, 23 April 1588.
155. BM Add. MS. 28, 360 ff. 171–7 and BN (Paris): *Manuscrits espagnols* no. 421 ff. 367v–72v contain Bono's proposals; AGS:GA 109/344, Alava to the king, 20 November 1581, Lisbon, shows what the captain-general of artillery thought of them, and includes the *consulta* with Philip's response on the dredging project. Bono was more successful with his diving-bell, which allowed two men to recover six anchors 'long lost and at great depth' from the Tagus at Lisbon. As a result the king gave him a privilege to use the device for ten years to extract pearls and sunken treasure, giving the crown a tenth of the proceeds: AGI:P 260i/10 contains the privilege granted on 27 February 1582, and documents relating to the testing of the device.
156. AGS:GA 266/175–177, the king to conde de Fuentes with a list of Altmann's inventions, 6 September 1589, Escorial.
157. *Ibid.*, 272/4, Francisco de Nieba to the king, 7 February 1589, Valladolid, with annotation.
158. *Ibid.*, 82/67, Juan Delgado to Mateo Vázquez, 25 February 1577, Madrid.
159. *Ibid.*, 64/310, draft of royal *cédula*, 1556.
160. AGI:P 259/45, royal *cédula*, 10 November 1573, Madrid.
161. Paul E. Hoffman, *The Spanish Crown and the Defence of the Caribbean (1535–1585)* (Baton Rouge and London, 1980), p. 88 states that this was done in 1556. For his expedition to the Magellan Straits in 1581 Sarmiento de Gamboa requested lead sheathing; MN:FN XX/50/480v, Sarmiento de Gamboa to the king, (n.d.).
162. MN:FN XXI/46/155–156, the Venetians' proposal to Charles V, 1555; *ibid.*, XXI/51/167–168, *cédula* of Philip II, 15 August 1556, Brussels.
163. The experiments and subsequent proceedings are documented in AGI:*Justicia, legajo* 1160. The case was heard within the *Casa de la Contratación* which functioned as a court of law in all disputes relating to navigation and trade with the Indies; appeals, as in this case, went to the Council of the Indies.
164. AGS:GA 77/164–166, description of Francisco de Contreras' proposal, with the king's comment, 1573.
165. AGS:E 1125/71; 82; 84; and 91, letters from Alvaro de Sande to the duke of Medinaceli, 4 June and 11 July 1560, Djerba; the duke to the king, 26 June 1560, Messina; and an account of the distillation at the fort by Pedro de Salzedo. The technique had already been suggested to Charles V by Blasco de Garay; *Codoin*, 81, p. 158. At Malta, during the rebuilding of the city after the great siege, the discovery of a spring of fresh water brought joy to the Grand Master – 'this is the best thing that has happened to me for a long time' – and an enthusiastic response from Philip: AGS:E 1131/102 and 109, Jehan de Valette to Philip II, 26 July and October 1566, Malta; *ibid.*, *ibid.*, 1131/102 records the king's reply.

166. J. Guilmartin, *Gunpowder and Galleys. Changing Technology and Mediterranean Warfare at Sea in the Sixteenth Century* (Cambridge, 1974), pp. 62–3.
167. The king to the duke of Medina Sidonia, 22 February 1588, in C. Fernández Duro, *La Armada Invencible*, vol. 1 (Madrid, 1884), p. 425.
168. IVDJ 45/67 records the experiments of August 1586, the views of Medina Sidonia and the *consulta* of the *Consejo de las Indias*.

# 4

## Producing the king's silver

The mining of precious metal was one of several means pursued by the crown to fund its military needs and its rising debts. The organisation which this required was a formidable undertaking. The French monarchy by the end of the sixteenth century managed its modest mineral resources through an administrator-general of mines; the Holy Roman Empire had detailed regulations for the intensive mining long established in certain of its lands; but nowhere was mining attempted on the scale of the Spanish monarch's operations. Much has been written about mining in the Indies and the treasure fleets; far less about the mines of the Peninsula itself. Yet at the beginning of Philip's reign they seemed to have equal importance for the solution of Castile's persistent financial crisis.

### *1 The rise and fall of Guadalcanal*

In the summer of 1555, while Philip was on his way to the Netherlands for the formal abdication of his father Charles V, Martín and Gonzálo Delgado discovered a rich silver mine at Guadalcanal, some twelve leagues north of Seville in the foothills of the remote Sierra Morena. The mine seemed to contain so much silver that it was soon being described as 'the richest that has ever been discovered in Christendom'.[1] To some it came as no surprise that Spain should yield so great a treasure. The chronicler Ambrosio de Morales took it as confirmation of what the ancients had written of the Peninsula's unequalled mineral wealth; they had marvelled at the 'auriferous' Tagus, the abundant silver mines worked by Hannibal in Cartagena, and the anchors of silver and gold made by the Phoenicians in Spain; and now in his own life-time had come this extraordinary discovery.[2] And those who had reflected on prevailing astrological beliefs that celestial bodies were constantly influencing the generation of precious metals within the earth had already become convinced that the constellations were especially favourable to Spain. When the astronomer Jerónimo Muñoz observed a comet over Elche he predicted, after consulting texts of the ninth-century Arab astronomer Albumazar, that rich mines were about to be discovered; Guadalcanal had proved him right.[3] And later, when Guadalcanal was under royal administration, astrological beliefs had some influence in encouraging expectations of further finds in the area. Diego Delgado, a clergyman of Madrid, professing knowledge of the secrets of planetary influences on the formation of metals, of why silver was generated more than gold in some earths, and the most propitious times of the year to look for seams, was appointed by the crown to make searches; he arrived at Guadalcanal in July 1556.[4]

When news of the discovery reached the crown, steps were immediately taken to secure the royal rights. In Castile, as in other parts of Europe, minerals were part of the monarch's regalia. By the late fourteenth-century law of Juan I, prospecting and mining were open to all, subject to the permission of the landowner, and on condition that the monarch was paid two thirds of the profit. Accordingly in October 1555 Juana, Philip's sister and princess regent, ordered guards to be put around the mine and all excavation and refining to be suspended so that the value of the ore could be determined by assay and claims established. All ore so far smelted was to be stamped with the letters 'REY' to indicate that royal duty was due on them; and soon 4000 ducats found in the possession of Martín Delgado was confiscated, a sum taken to represent less than a quarter of the value of the 15 *quintales* of silver he had already produced without payment of royal duty. Later that month the astonishing results of assay by a local smelter had been received; it was reported that one *quintal* of ore contained one *arroba* of silver, a proportion 50 times greater than was usually found in the more common argentiferous lead ore. The crown needed no further encouragement, and after consultation with the Council of Finance, Agustín de Zárate, recently returned from Peru where he had witnessed silver mining, was appointed to direct operations at Guadalcanal, leaving the claims in abeyance. He was instructed to employ German miners already at work on other sites in Spain and to make use of their equipment. The crown came to rely on Italians for military engineering; but for mining expertise northerners had the most to offer and since the early years of Charles V's reign, men from Limburg, the duchy of Lorraine, Saxony and the county of Tyrol had been brought to Spain to exploit mineral resources, especially the mercury mine at Almadén. In 1553 Prince Philip had sanctioned a comprehensive ten-year contract with Hans Schedler to import skilled Germans 'who know how to do things cheaper than is usually done here' to operate mines throughout the Peninsula. Schedler would pay the crown 166 250 *maravedís* a year and duties on metals extracted; what remained was for Schedler and his workforce. Now many of the Germans were to be transferred to Guadalcanal to secure efficient and economical production.

Two other priorities were made clear in Zárate's instructions. He must be vigilant about security: all ore was to be kept under lock and key, and all men entering and leaving the pits were to be kept under surveillance to prevent concealment of ore in their clothing and other theft. He must also guarantee steady supplies of wood to the site, indispensable for fuelling the furnaces and for supporting the shafts with timber props.[5] Security and wood supplies were to present recurring difficulties for the administrators of Guadalcanal.

In November 1555, three weeks after receiving his instructions, Zárate had made preliminary investigations on the site and was preparing to send a report to the princess regent. He regretted that up to his arrival some 10 000 ducats of silver had been stolen as news of the discovery attracted pilferers from 20 leagues around. He issued proclamations in Guadalcanal and Seville that all unstamped silver would be seized; he also had a stamp made with the royal arms, which was kept in a locked chest locked within a second chest, 'because everything depends on the safekeeping of this die, and it has to be guarded as well as it is in the Indies'. Already there was a dispute with the town council of Guadalcanal over cutting wood. Zárate was in no mood to be delayed. Heavy continuous rain since his arrival had brought fears that the mines would collapse and attempts to introduce props had been prevented by falls of earth; the rain had also halted smelting because none of the buildings yet had tiled roofs. Ignoring the town council's refusals Zárate told the princess: 'I did the cutting without anyone daring to resist'. For the moment a stock of wood had been secured for the mine. Other bad news indicated the numerous parties who were pressing their claims against the crown for the silver treasure. But the rest of the report revealed Zárate's excitement and brought delight to the princess. Here was ore which yielded 'between one third and one fifth of its weight in fine silver, which is something that has scarcely been seen in Potosí'; the pits 'will produce such wealth as has never before been seen in these realms; they bear all the signs of perpetual wealth'.[6]

In her letters to him over the next few months the princess made the crown's policy on the mine quite clear. How soon could he provide 60 000 ducats for the fort at La Goleta? Silver to this value must be sent 'as quickly as possible' to Seville for coining so that it could be given to Diego de Cazalla, paymaster of the armada, for shipment to that African fortress. And in February 1556 she informed Zárate that 40 000 ducats were urgently needed to pay soldiers on the French border at Fuenterrabía and San Sebastián, and 'you are to provide this as soon as you can'. Zárate was told to make sure that the German workers were being well treated because 'they are so essential'; they and the local workers were to 'extract all they can as quickly as possible'. The opening of more pits with the help of drainage equipment was to be investigated because 'this could give quicker returns to the treasury'. And he was ordered to take care that 'silver is collected from washings, second washings, sweepings, dross and other small quantities so that nothing is lost'.[7] Opportunities for fraud in the various stages of silver processing were minutely scrutinised; and after instructions from the princess to the officials of the *Casa de la Contratación*, where the silver was to be delivered for minting in Seville, two strong chests, reinforced with iron and each provided with three iron locks, arrived at Guadalcanal for the transport of bullion. The princess told Zárate 'expectations are so great, and all should be done with the greatest speed'. She kept the king informed of latest developments at the mine, reporting in February 1556 that 'so far the yield has been 20 000 ducats' which was given to Andrea Doria, captain-general of the sea, for payment of galleys; but more was owing to him and to the soldiers at Perpiñán, Rosellón and the Balearics, in all a debt of about 100 000 ducats 'without taking account of expenses this year, and there is nowhere from which this may be supplied'.[8] Nowhere immediately, that is, apart from what Guadalcanal might yield.

As news of the mine spread the princess complained of harassment by the crown's numerous creditors; she wrote to Zárate: 'Therefore do not tell everyone about the yield, but only those of the Council of Finance'.[9] The archbishop of Toledo thought the creditors should be kept waiting with promises of payment. He wrote to the king, once his pupil, that 'God at the beginning of your reign has revealed in these kingdoms of Spain more treasure than was ever disclosed to the Romans when they were here'. He advised that instead of using this wealth to pay off debts, it should be accumulated in the alcázar of Toledo. And when neighbouring princes learned of this monetary power 'their wings would be clipped, they would not dare to be as bold as they now are, and they would send ambassadors seeking friendship just as happened to King Solomon by virtue of his great treasure'.[10] That policy was not adopted and orders continued to be dispatched to Zárate for silver to meet urgent military needs, giving them priority over any repayments of debts to Genoese bankers. In 1556 Guadalcanal provided 22 000 ducats for soldiers on the French frontier; 9 000 ducats for artillery and food for the men at Melilla; 23 000 ducats for the armada sent to relieve Oran from the besieging Barbary allies of the Turks and a further 7 500 ducats for its defence. And in July just when attempts were being made to improve the processing of ore by the use of thermally efficient reverberatory furnaces, the princess cancelled the introduction of the modified techniques, which would take time to implement, and ordered a continuation of the existing process, because news had come of the siege of Oran and 'with each hour of delay much is at stake'.[11] In 1559 silver from the mine to the value of 100 000 ducats was spent on Andrea Doria's galleys.

One of the earliest descriptions of the site at Guadalcanal comes from a document of February 1556 written by Alonso de Chaves, the royal cosmographer who had been sent from the *Casa* in Seville to paint a coloured map of the mineshafts. This means of facilitating mining operations was again used by the crown five years later after the mine had expanded; Juan de Orihuela, an artist from Granada, was commissioned to paint a plan of the subterranean works to guide the foremen and workers, and the canvas, over 15 feet long, was hung in the church on the site for all to see. Both canvases have been lost, but the key which Chaves prepared to accompany his illustration has survived and provides some interesting new information.[12] He found that no fewer than 102 pits had already been dug, mostly shallow, no more than a few feet, but some were already ten *estados* (about 60 feet) deep. 33 of these were known as 'the pits of the great company' because they had been jointly registered by numerous inhabitants of Azuaga, a *pueblo* four leagues to the north. A group of thatched huts housed the German workers; separated from these were other huts for workers from neighbouring *pueblos*; another hut, its entrance guarded by a dog, served as a prison. Ore lifted to the surface was stored in a tiled hut, around which timber brought to the site was unloaded. The ore was washed in wooden boxes in a ditch fed with water from the small river nearby. The silver ore was then ground, mixed with litharge flux (a lead compound added to facilitate smelting) and taken for smelting to one of five furnaces. Finally the smelted product was refined in seven furnaces. Work continued day and night, stopping only on holy days. This was the same smelting process which the Germans had employed so successfully in Saxony, Bohemia and the Tyrol, making those central European regions the West's greatest silver producers in the early sixteenth century; and it continued into the twentieth century to be the basis of one method of silver production.

More details of operations at Guadalcanal were given by Zárate in one of his reports to the princess in February 1556.[13] Work was organised in eight-hour shifts, during which a team of five men went to each of the 11 pits then being worked. While two men excavated with chisels and picks, three others operated a hoist using ropes and buckets which brought ore to the surface and removed water from the pits. After washing and crushing the ore was loaded into furnaces in quantities of 30 *arrobas* and mixed with 16 *arrobas* of litharge flux. Fuelled with charcoal and firewood the furnaces were opened at the right time and molten silver–lead alloy flowed into a pit, leaving intractable solid impurities, the gangue and dross, in the furnace. After cooling, the silver–lead in the form of slabs was loaded on mules and taken for refining to furnaces one league away – the reason for this inconvenient separation of operations is not made clear, but the princess soon suggested that refining and smelting should be done on the same site. 165 men were then working in the pits, earning 3 *reales* a day; they were mostly German and said to be honest in contrast to the rest,

Moriscos and other Spaniards from the area. There were also German smelters, German carpenters to shore the shafts, and Spanish refiners. The daily running costs were 40 593 *maravedís*: 1800 for oak timber; 2496 to pay Zárate and nine other officials; 3700 for charcoal fuel; 6000 for litharge flux; but the workers' wages accounted for most of this expense, 26 597 *maravedís* a day. In addition payments at unpredictable intervals had to be made for hide buckets for drainage, panniers for collecting ore, ropes and oil – by the end of 1558 two and a half *arrobas* of oil were needed every 24 hours to light the pits, though the cost here was small, no more than 600 *maravedís* a day (in 1568).

The crown from the start kept a close watch on expenses and had to deal with artificially high prices of litharge caused by individuals hoarding stocks in anticipation of the new demand at Guadalcanal.[14] Profiteering at the expense of the crown's undertaking was also attempted by the town council of Guadalcanal which introduced a levy on litharge, charcoal and wood. 'You know these things are needed', the princess wrote to the council, 'and I order you not to make them expensive'.[15] In addition bargees were said to be overcharging the crown in ferrying supplies from Seville.

But the crown was in no hurry to economise on labour costs; on the contrary efforts were being made to import more of the prized German workers in large numbers to boost production. Zárate wrote to the king in March 1556 asking for arrangements to be made to send 200–300 Germans because 'around the town there is more silver than in Peru or New Spain'.[16] Similar recommendations came to the king from the Council of Finance, and from the princess who also advised caution to ensure that Lutherans were not included in any imported German workforce. In May Philip wrote from Brussels to the banker Anton Fugger in Augsburg asking him to find 200 Germans who would be willing to go to Spain to assist silver production, giving his assurance that they would be well treated.[17] The result of this request is not recorded; but the workforce at Guadalcanal continued to grow: there were around 200 in February 1556, over 300 in June – 'almost a small town' Zárate said – and 1000 in the following January.[18] On several occasions the princess instructed Zárate to mingle the Germans with the Spanish workers both for the encouragement of honesty and the dissemination of technical knowledge.

Other measures were taken by the crown to maximise production at Guadalcanal. Furnaces were prepared a week in advance, fed with charcoal and heated up for use just when they were needed, so that no time would be lost. And to increase the working days at the mine, the princess wrote to the archbishop of Seville and the prior of San Marcos de León to permit work on feast-days; this was granted in exchange for the payment of alms. But the crown was not yet satisfied that everything possible was being done in the interests of efficiency and economy, and looked for additional expert advice. In April 1556 the princess appointed Francisco de Mendoza to investigate how output at Guadalcanal might be increased. The son of a former viceroy of New Spain and Peru, Mendoza was said to have experience of mining in both of these lands, but exactly what his involvement had been is unclear. And later the same year Hans von Estenberger came to Spain, commissioned to scrutinise operations at the silver mine to see if costs could be reduced.

Mendoza's arrival was of greater consequence, and admiration for his firm and energetic presence at Guadalcanal soon led to his appointment in 1557 as the crown's first administrator-general of mines of the entire Peninsula. Although Zárate was regarded as a competent and honest administrator, Mendoza found him inflexible and unwilling to believe that any improvements could be made. But the crown obliged Zárate to observe Mendoza's recommendations. The potential source of conflict did not last for long; within a few months Zárate was sick and had to be relieved of his duties.

Mendoza was required to work closely with the *Consejo de Hacienda* (Council of Finance). This advisory council, created in 1523 to take charge of Castile's finances, became increasingly active in mining affairs during Philip II's reign; a small body which met daily, it consisted of a president, three councillors and three to four *contadores mayores*, supervisors of accounts. Mendoza's first moves at Guadalcanal were to recruit extra labour and introduce draconian measures to eliminate losses from theft. In May 1556 an assembly of all those on the site heard the proclamation of severe penalties – a hundred lashes and years of galley service – for thefts of silver; restrictions on residence on the site removed taverners and all others who were not involved in mining; a dusk-to-dawn curfew prohibited outsiders passing signs indicating the site's perimeter, and unauthorised departures by workers from the site during these hours were punishable by a hundred lashes; all weapons were forbidden on the site unless special licence had been obtained.[19]

Mendoza also reported that open-cast working, which the princess had asked him to consider to facilitate production, was impracticable at Guadalcanal. Nor was there any way to reduce costs by the use of water power; there was hardly any water available, and the more expensive animal power would have to continue for operating grinding mills and the bellows which provided the draught for the furnaces. But he was convinced that fuel costs could be reduced by introducing reverberatory furnaces of the type he had seen in New Spain, and arrangements began to be made for their construction. And to see how the high cost of labour might be reduced he calculated that 2500 *maravedís* a year could be saved for each negro slave employed instead of Spanish waged workers. The saving was not greater than about 12 per cent because although slaves were not paid wages the cost of feeding and clothing them had to be borne, and allowance had also to be made for 'the usual five per cent who die'. Mendoza wanted the slaves to come from Cape Verde and thought they could be put to work in excavation, drainage or operating bellows.[20] Slaves were used at Guadalcanal from at least 1559; there were 60 on the site in 1564; and on 29 October 1565 descriptions were made public in the market place of Llerena of two slaves who had escaped from the mine on the previous night.[21]

Despite the crown's close attention to the management of Guadalcanal in its insatiable demand for silver, all attempts to achieve continuous production were frustrated by various obstacles. The climate frequently hindered operations. In the winter heavy rain caused subsidence and flooding of pits; in the summer months there was usually no smelting or refining because the workers could not tolerate the intense heat; and in July 1556 Zárate was waiting for the rains to come because the river had dried up, preventing all washings of ore, and all that could be done was to search the river bed for silver which might have been lost in previous washings. The interruptions caused by illness also worried the crown. The princess urged Mendoza to see that 'men who become ill or die, especially the smelters and refiners, are replaced so that no part of the operation ceases'. And when in September 1556 she was seeking a new administrator at the mines she rejected Philip's ailing nominee as unsuitable because 'a person of robust physique is needed to withstand the dangerous fumes of smelting and refining, which afflict many workers and have even killed them'.[22] In January 1566 the administration was inconvenienced by the loss of a much-needed official: a guard had to be discharged because of 'the continuous illness he and his wife suffered on this unhealthy site from the fumes and noxious vapours'.[23] In August 1556 Zárate reported that he, Mendoza, and almost all of the Germans were ill; consequently he was forced to employ 40 Moriscos from Hornachos,[24] a town ten leagues to the north, whose population, almost entirely Morisco, had skills in mining and refining. Their Morisco foreman, Francisco Blanco, would become one of the most valued workers at Guadalcanal. On other occasions recruitment of local labour proved difficult because of the demands of harvesting.

Above all, the site was vulnerable to any failure in supplies, which repeatedly occurred. And for supplies of all things the administrators depended on the arrival of carts and mules carrying materials and equipment from a wide surrounding area. The muleteer who arrived at the mine on 20 May 1559 brought a basket of 17 chisels; 6 *arrobas* of tallow candles for lighting the pits, and another 8 pounds of wax candles for mass; 50 esparto ropes; 10 pounds of thread; 4 reams of paper; and a brass weight of one mark.[25] No difficulties were experienced on that occasion. But from time to time the town council of Guadalcanal, claiming jurisdiction over the mining site one league away, had interfered with these baggage trains and a royal cedula had to be sent with instructions not to obstruct the traffic of goods.[26] And later during the Granada uprising when all of the droves of the local muleteers had been commandeered for the service of the king's soldiers, the officials of the mine sent a desperate appeal to the king to countermand this because the resulting shortages of charcoal threatened to bring all smelting to a halt.[27]

Supplies of charcoal and wood depended on the good will and obedience of other authorities. Large cities needed these things for their own purposes and by one means or another the crown tried to secure their agreement to the demands of a new competitor. The princess sent orders to Córdoba demanding permission to take charcoal to Guadalcanal for smelting and refining even though existing municipal regulations prohibited this until the city's fuel requirements had been satisfied; the royal letter could only add 'provided no harm is done to the city'. With Seville the princess used a more devious approach. The city relied for its timber on the large oak forest of Constantina, a small town under its jurisdiction and close to Guadalcanal. This was the convenient source of structural timber and fuel that the crown had been looking for; the princess wrote to Seville for permission to take 200 cart-loads of oak from the forest and, revealing that this recognition of Seville's jurisdiction was no more than lip-service, she simultaneously instructed Constantina to supply the wood even if Seville withheld permission.[28] There were times when Constantina did not comply, obstructing the mine's contractors who had been sent to manufacture charcoal, or diverting its wood supplies to Seville. But for the most part Constantina seems to have supplied Guadalcanal regularly and soon became the principal source of the mine's fuel and timber; by 1558 a team of eleven carters provided the transport of this indispensable material from the forest to the mine four leagues away, a journey which must have been slow because the route lay in the sierra.

Individuals as well as towns interfered with the provision of the mine. The jurisdiction of much of the land around the mines of Guadalcanal was in the hands of the Order of Santiago, the wealthiest of the Peninsula's medieval crusading military–religious Orders. The power of the Orders had been curbed by the crown through the securing of their Masterships by monarchs since Ferdinand the Catholic. But in Philip's reign audacious members of the Order could still challenge royal authority. When in 1579 a royal official went from the mine in search of wheat and barley to feed the workers and the mules which provided the power on the site, he was arrested by the local governor of the Order, his letter of royal commission confiscated and his wand of office seized. A request from the mine's administration to return the royal warrant had been ignored and consequent shortages of food had brought work at Guadalcanal to a stop. The king now issued a stern letter blaming the governor for lost silver production and warning him not to wait for another order to cease interfering with the king's business, or face a fine of 1000 ducats and other punishment 'at our pleasure'.[29]

Despite the difficulties Guadalcanal produced a considerable quantity of silver for the crown in the late 1550s to the great satisfaction of the princess regent and the king. In July 1556 official accounts showed that in the eight months since operations began silver worth more than 102 million *maravedís* had been produced. This is to be compared with contemporary production in the Indies. For the year 1555 silver production at Potosí has been estimated at 265 893 marks;[30] for November 1555–July 1556 the yield at Guadalcanal was equivalent to 44 036 marks.[31] And some indication of the importance of this revenue for the treasury can be judged by comparing it with bullion receipts, both silver and gold, from the Indies, which amounted to just under 443 million *maravedís* in 1555, most of this going to merchants and other individuals and about 30 per cent to the crown.[32]

But while the crown was spending the proceeds from Guadalcanal it knew that sooner or later a high price would have to be paid for this windfall. Settlements with the discoverers and other claimants were yet to be concluded; and until then no one was in a position to make a proper calculation of the royal profits. The matter was more far reaching than that. After discussions with Mendoza, the Council of Finance advised the king that while the discovery at Guadalcanal had given a spur to prospecting in other places, men were reluctant to reveal their discoveries until they could see what settlement was made with the discoverers there; a general declaration by the crown might be necessary to clarify the conditions for prospecting and the rights of discoverers. Mendoza thought this 'the most important matter for discussion in Spain today'; he urged the Council of Finance to formulate a proposal and begged the king to send his decision from the Netherlands so that regulations for mining in Spain could be published. He told Philip of the numerous rumours of rich mines and that even if many would fail to be confirmed, he was convinced that others would provide wealth 'no less than those of Peru'.[33]

In February 1557 Philip informed the Council of Finance that he preferred a policy of complete freedom to prospect but wanted to incorporate all the Peninsula's mines within the crown, though Guadalcanal appeared to be a special case and he asked the Council for further advice on it.[34] During that year legal experts were busy offering advice to the crown on how best to deal with the outstanding claims; their deliberations reveal the particularly complicated dispute over Guadalcanal.[35] There was Juan Diez de Vivar, owner of the land on which the mine was discovered; he claimed rights to one third of the profits on the grounds that he had not given permission to the discoverers to prospect or excavate mines on his land. The two discoverers were in dispute over this third, and one of them, Martín Delgado, now deceased, was alleged to have promised smaller shares to others like Thomé Hidalgo who claimed that he was the first to smelt the ore and demonstrate its high quality. A resident of Trujillo, Diego de Vargas Carvajal, also came forward with claims based on a concession given by the Catholic Monarchs to his father of the royal duty on all mines discovered within the diocese of Badajoz and three leagues around; he supposed the mines of Guadalcanal to be located within this area. And there were Hans Schedler and his German associates with whom the crown had signed the contract of 1553 authorising them to operate mines in lands which certainly included the Guadalcanal area for a fixed annual payment; and now that the opportunity of exploiting this richest of mines had been taken from them they were claiming tens of thousands of ducats compensation from the crown for what they had already invested without profit.

The lawyers argued that the monarch was the true owner of mines in Spain and all concessions to individuals by previous monarchs were subject to his confirmation; and consequently that if the king chose to take the mines of Guadalcanal for himself he would have justice on his side. Nor, they maintained, were the discoverers entitled by law to any share of the mine or its profit, except by the grace and favour of the king. And Venero, the king's *fiscal* (crown lawyer), regarded the claimant's demands from previous monarchs' acts and concessions to be unjustifiable, because 'it is a principle in law that the royal grant of privilege extends only to what was conceived when it was bestowed and not to what was inconceivable. And since the wealth of this mine was inconceivable until now, nor was conceived in all the ancient chronicles Roman or Castilian', past monarchs did not know what they were conceding and their grants were therefore invalid. He thought the discoverers of Guadalcanal should receive no more than one hundredth of the profits.[36] It was said that if Juan I had known of the wealth of Guadalcanal he would never have enacted a law giving discoverers one third of the profits; so large a share was relative to the value of silver mines then known, which was 'very little'.[37] But if past grants and concessions could be put aside, the crown's contract of 1553 with the Germans could not be so easily dismissed, and the lawyers recommended that they be compensated for the risk of their investment and the costs of operations in the years preceding the discovery of Guadalcanal. Eventually in 1561, after protracted negotiations, Schedler and two German associates agreed to accept 33 500 ducats each in crown annuities, and in addition Philip waived his right to 500 ducats in duty for ore which they had extracted; in renouncing all their claims to Guadalcanal's yield they made it clear that the expense of lengthy legal proceedings and the doubtful outcome of the suit they had brought against the crown had forced them to settle. But the heirs of Martín Delgado, the discoverer, held out longer; in 1564 the Council of Finance was still trying to persuade them to settle for one sixth of the mine's profits, a sum amounting to 101 million *maravedís*, out of which they themselves would be required to compensate the landowner and other claimants, 'leaving your Majesty free of disputes'.[38] This they eventually agreed to do, but the dispute was not settled until 1568–9 and the price paid by the crown to the heirs of Martín Delgado may have been as high as 100 000 ducats.[39]

While these disputes were dragging on, the king issued comprehensive regulations on the mining of precious metals (January 1559 and March 1563; others were to follow in August 1584) which were to control mining in Spain for two centuries. All silver, gold (and mercury) mines in the Peninsula were declared incorporated within the crown and the area within one league radius of the mines of Guadalcanal declared a royal preserve in which private prospecting was prohibited; the same applied to sites at Aracena, Cazalla and Galaroza, other silver mines recently discovered in the Sierra Morena. Everywhere else prospecting and mining would be open to all including foreigners; the only exceptions later made to this general invitation to enterprise were the royal mining officials, who were forbidden to hold mining interests, and the clergy, whose involvement in mining, regarded by the king as 'indecent', led him to instruct prelates to punish offenders.[40] The crown's intention was to stimulate mining in partnership with its subjects for the benefit of all; therefore it had revoked all previous grants which had put mines in the hands of inactive individuals – they would be compensated but not permitted to remain in passive possession. Now within a few months of their discovery silver or gold mines had to be registered, shafts sunk to a depth of three *estados* and at least four persons employed to begin operations; otherwise the mine would be confiscated. As for the usufruct, the third of the profits prescribed by Juan I's act was moderated to allow for rich mines like Guadalcanal by imposing thresholds on profits – once these reached 100 000 ducats the operator's share would fall to one quarter, and then to one fifth for profits over 200 000 ducats, the crown's share rising accordingly. In the ordinances of 1563 a similar effect was achieved by introducing a sliding scale based on the quality of silver ore: for ores containing up to three marks of silver per *quintal* the crown was to be paid a duty of one quarter; rising to one half when ores processed yielded six marks of silver per *quintal*.

It is clear from the ordinances of 1563 that the crown envisaged Guadalcanal as the hub of a peninsular network of mines. The royal officials at Guadalcanal were given the responsibility of keeping a register of all mines discovered and sending bi-monthly reports to the *contadores mayores* of the Council of Finance on the state of mines and their yield. The same officials were given powers to allocate funds for mining expenses. And the regulations for refining – all important for the crown to control because this was the stage at which the royal duty was taken – required the erection of a royal refinery in every silver-mining site where all silver–lead smelted had to be refined by refiners appointed by Guadalcanal's officials.[41] The itinerant administrator-general was to ensure that operators were shoring and propping their mines adequately and constructing furnaces (plate 8).

One of the reasons for Philip's return to Spain from the Netherlands in September 1559 was the urgent need to improve the state of the royal finances in Castile. For a while he continued to be encouraged by Guadalcanal's yield and approval was given for the creation of a central fund in the *Casa de la Contratación*, fed in part by the silver mine; the failure of this to develop into the monarchy's commercial bank has been blamed on the officials of the *Casa*.[42] Mendoza's reports from Guadalcanal had continued to be optimistic. But in January 1563 came the first signs that the bonanza was coming to an end; it must have been a bitter disappointment and it came at a time of a sharp dip in silver receipts from the Indies, and when the crown's military expenditure was about to rise enormously – after 1566 there was war against the Turk, the Morisco rebellion and the draining revolt of the Netherlands.

The king received a report from his officials in Guadalcanal that

Pregmatica de su Magestad, por la
qual reuoca las mercedes de minas, y declara la parte que
han de hauer los descubridores dellas, y la orden
que se ha de tener en registrarlas
y beneficiarlas.

1559.

¶Con priuilegio: Tassado a cinco marauedis el pliego.

Plate 8 Philip II's pragmatic of 1559 incorporating the mines of Spain within the crown. With the modifications of later ordinances (1563 and 1584), these regulations continued to control mining in the Peninsula in the early nineteenth century. (AGS:DC 46/22).

little silver ore was being extracted and none at all for a whole month;[43] in sharp contrast to months such as June 1557, when 600 *quintales* of ore was extracted in a single week. Faced with an annual salaries and wages bill of well over four million *maravedís*, the mine's officials asked the king if workers were to be dismissed and salaries reduced. The king replied that superfluous employees were to be dismissed and those retained given a reduced wage in accordance with their lighter burden. There were too many guards and one of the two chaplains must go; savings could also be made by officials serving in more than one capacity – the treasurer was soon instructed to act also as the justice of the mine. And as a result of the diminishing returns the king removed the prohibition on prospecting within the royal preserve; in August he approved a contract with Jerónimo de Salamanca, a prospector from Burgos, to search for new mines in the Guadalcanal area, as usual leaving all expenses of operations to the prospector and requiring a share of any profits. In the following March the king announced drastic measures; production had remained negligible and even after the dismissals and reduction in salaries, expenses were still rising and exceeded the yield from the mine.[44] Philip therefore ordered the immediate dismissal of all salaried officials except for the treasurer, who was given custody of ore, tools, machinery and slaves; the *contador*, retained to keep the accounts and 'look after all other business of the mine'; and the German assayer Lucas who was made curator of the mine chiefly to prevent fire, flooding or collapse. The king instructed the *Casa* to release 3000 ducats to pay off dismissed employees. But this was not an order to stop mining at Guadalcanal; the king wanted excavation, smelting and refining to continue, employing a minimum of essential workers with moderate wages.

Apart from excavation there was work to be done on the stores of ore and the processing of residues. And for this the king was still willing to invest large sums of money. This is evident from his response to the officials' request in September 1566 for a reserve of ready money at the site instead of the periodic collection of funds from the *Casa* at extended intervals when refined silver was taken there. There was under 200 ducats in the site's chest and a lot more than this would be needed to pay for the indispensable litharge flux, charcoal fuel, ropes, wheat and barley, all of which required immediate cash payments; also ready money was needed so that workers were not kept waiting for their wages. The officials asked for 12 000 ducats and were given 10 000.[45] But in the early 1570s there were continuing complaints of delays in receiving money; in 1574 6000 ducats assigned to the mine were diverted from the *Casa* to pay the troops in Flanders.

Another sign of the lowering of the crown's expectations from Guadalcanal may be detectable in its plan to sell the town of Constantina. The sale of towns belonging to the crown was frequently resorted to by Philip to supply much-needed income. Constantina was no ordinary town, and representatives of its 1200 residents who were reluctant to see it alienated from the crown, reminded the king of its valuable timber which had not only assisted the Catholic Monarchs' reconquest of Granada through the construction of gun carriages, but had in recent years supplied the mine of Guadalcanal with an estimated 80 000 cart-loads of wood. And although Guadalcanal was in decline, the town's forests would continue to serve as the principal source of supply for any future discoveries of mines in the region. The king was warned that if he sold the town the timber might be more difficult to obtain or at a much higher price. In the 1550s and early 1560s the sale of this asset would have been unthinkable; but now it seemed less important and by September 1565 Don Fadrique Enriquez de Ribera, president of the royal Council of Orders, became the new señor of Constantina.[46]

The king seems to have been less carried away than some of his administrators by exaggerated visions of unlimited silver treasure; and from time to time he had inquired more soberly about the mine's prospects. But the fall in production had been sudden and Philip approved the Council of Finance's recommendation to send someone to the site 'to investigate why mines which had given so much wealth were ceasing to be profitable'.[47] By March 1570 the abandonment of Guadalcanal was being seriously considered in Madrid. And then in July came news of the discovery of a seam alleged to be as rich as that found at the start.[48] The mine's officials requested the restoration of the workforce, the return of the guards and, as clear a sign as any of revived expectations, the recall of the assistant priest to cater for the needs of an enlarged working community. In almost a repeat of what had occurred 15 years before, Zárate, the mine's first administrator, was appointed by the king to return to the site he knew so well to investigate the new strike;

he was soon once again sending back excited reports: the results of an assay showing that a *quintal* of ore from the seam contained 163 marks of silver (over 80 per cent) which 'has never been seen in our time'.[49]

But Zárate found that conditions at the mine had changed. The cost of provisions, especially wheat, had risen considerably; consequently the guards who had been paid three *reales* a day in 1556 now had to be given six *reales*. The most important difference however was the greater depth of the pits; the new seam had been discovered 80 *estados* (over 500 feet) below the surface and its exploitation was far more demanding than previous mining on the site. The heat forced the miners to enter the pits naked, and lack of ventilation made both respiration and the lighting of candles difficult. Was it worthwhile to proceed? Men on the site with experience of mining in Saxony, Lorraine and the Tyrol were approached for advice.[50] They said that in those lands any such indications of mineral wealth would lead to immediate investment even if the pits were deeper than these and the ore one tenth of the richness of that found in the new seam. This encouraging assessment was qualified by their statement that there could be no guarantees in advance that the promising signs of wealth would materialise; experience alone could determine this and that meant the construction of a ventilation shaft to enable work to continue on the seam.

The crown had taken advice from expert foreign miners and there seemed to be grounds for persevering. Work began on the excavation of a ventilation shaft. This was supervised by Francisco Blanco, the Morisco who had discovered the new seam; he had been enticed back to the mine by the promise of repeal of a sentence of exile from Hornachos, imposed 20 years before by the Inquisition. Now all hopes were placed on his skill and Zárate kept the king regularly informed of his progress. The task was arduous because of the hardness of the rock; this had always hindered excavation at Guadalcanal. At a time when blasting techniques employing drills and explosives had not yet been introduced in mining, excavation was bound to be slow. Zárate told the king that the best miner could not dig more than a bucket of rock a day. 120 iron picks were attacking the rock day and night, but the tools were being worn away; in places 'the picks had as much effect on the rock as on an anvil'.[51] Flooding was another obstacle. Master

Hanz, a German craftsman brought from the mercury mine at Almadén, had installed a drainage engine in which mule power was applied to a hoist of chains, lifting water to the surface in hide buckets. But the chains, hundreds of feet long, often snapped under their weight, causing frustrating delays and expense. Zárate wanted stronger chains to be made in the iron works of Vizcaya, sending them by sea in units of six feet for assembly at the mine.[52]

In March 1571 after four months of toil Zárate informed the king that the ventilation shaft 'which has held us in suspense for so long' was at last completed. It brought 'a great rush of wind' and miners who had been working naked now had to wear clothes because of the chill. There was no immediate gain and in May the *contadores mayores* ordered Zárate back to Madrid; but, convinced that great riches were within sight, he refused to leave. The king urged the mine's officials to 'secure the fruit with all possible diligence' and said 'you can understand how important this could be for our needs'. He ordered an investigation of how drainage might be achieved more cheaply and easily, the replacement of worn-out slaves, and the increase of the workforce if this seemed necessary.[53] There was to be no second bonanza; but over the next three years silver production from the new seam brought around 150 000 ducats to the royal treasury.[54]

The rising yield was not sustained. During 1574–5 production was again slight and the abandonment of the site was again under consideration. The officials' reports presented an image of exhaustion. The ore was running out; the tools were good only for scrap iron; the mine's 20 slaves were exhausted or decrepit; the 14 mules were so worn-out that expensive manual labour had to be substituted for operating the drainage buckets and the bellows for smelting. As to whether all operations should come to an end, they advised the king that this was difficult to judge because no one could be certain of what lay under the earth; Blanco had assured them that 'there will be ore, but when and how much remains to be seen'.[55] The king was not satisfied with this and pressed the officials for a clear recommendation, asking also if the funds needed to maintain operations could come from the mine's silver. The officials replied that there was no silver and the money – 15 000 ducats a year were requested – would have to come from the king. The mine, once viewed as the great remedy for the crown's debts had become just another drain on the royal treasury.

After waiting six months for the crown's decision the officials wrote in May 1576 that the matter had been decided for them by the fracture of the drainage engine and the flooding of the pits. Materials and equipment which could be retrieved were quickly taken from the pits; the workers paid and dismissed; all mining had come to a stop. The officials recommended the sale of the mine's stock, advertising this in localities within an eight-league radius, especially with a view to interesting the Fuggers, lessees of the mercury mine at Almadén. But they wanted the most valuable piece of equipment to be kept: the giant wheel of the drainage engine, which cost over 4 000 ducats to make, was in good condition and might serve as a prime mover for another engine. The church on the site should also be conserved to commemorate the great wealth and prosperity which God had given and in whose name all had been done.[56]

This was the end of royal administration at Guadalcanal. The Council of Finance sent instructions for the sale of the slaves – if a good price could not be obtained they were to be sent to Seville because they were needed in the Indies. A complete inventory was to be prepared and the mine's account books brought to Madrid. The inventory portrayed a vivid image of mining operations over two decades: hundreds of picks and pickaxes; dozens of hammers; ladders, ropes and pulleys; numerous chains, some used for the drainage engine, some for measuring the depth of pits, others smaller for shackling the negro slaves; bells for use in the pits; furnaces and crushing mills; cow hides used for bellows; scales, some for assay, some for weighing refined silver, others for weighing the bread given to the slaves; 24 chests for carrying the silver to Seville; and, from the mine's pharmacy, alembics and vessels for preparing medicines, and a portrait of the Virgin.[57]

Soon there was no one left on the site but the *contador* and the priest, who, reluctant to desert the church where many workers and slaves were buried, offered to remain as a hermit. But this was not the end of operations at Guadalcanal. In the 1580s the crown leased the site to Janus Julius and Ferdinand Favolus, physicians of Arras and Cologne, in exchange for a share of the profit; but soon they were accused of breaking their contract by neglecting drainage. And soon after this Andrés de Tolosa was attempting to produce silver there; he was convicted on 88 charges of theft and embezzlement. In 1581 when the Council of Finance was searching for revenue to pay the troops in Flanders attention turned to the heaps of residues at Guadalcanal and plans made for recovering their silver content by a process of amalgamation.[58] But this too proved unsuccessful; there is no mention of Guadalcanal silver in the statements of crown income after 1580.[59] Philip still entertained hopes in July 1597 when he appointed Gerónimo de Ayanz his new administrator-general of mines, instructing him to visit Guadalcanal with a view to planning a drainage project, and 'if this should be too expensive or difficult you are to see if the vein can be discovered by another approach; since it was so rich there is reason to believe there is some outcrop nearby'.[60] But in 1602–3 Ayanz had decided that Guadalcanal should not be revived; drainage would cost a 'huge sum', structural support would require timber costing 300 000 ducats, and all this when 'no one knew what existed there'. Philip IV eventually gave the mines in perpetuity to the Fuggers for their financial services to the crown, requiring only the payment of royal duty; but flooding prevented extraction and the mines were abandoned until the eighteenth century.

The peak years of peninsular silver production had been 1556–9 when an average of 67 500 marks of silver were produced. The best year was 1556 with 74 715 marks; but after 1560 annual production never reached 20 000 marks; and in 1575 it had fallen to 2405 marks.[61] When Guadalcanal was at the height of its production, official estimates indicated that for the period 1555–63 the crown's profit was over 606 million *maravedís* (production of silver valued at 740 million *maravedís*; expenses at 134 million).[62] But still to come were heavy expenses for compensation and the sharp decline in yield.

## 2 *Silver from the Indies*

So long as Guadalcanal was productive the crown must have derived additional satisfaction from its location. The treasure was close at hand, within the Peninsula, and royal control thereby facilitated. Zárate, its first royal administrator, had been given his instructions on 29 October 1555 in Valladolid; on 6 November he had arrived at Guadalcanal, dispatched his first report on the 20th and, after it had been received and discussed, the princess regent had sent her reply on 18 December. Correspondence occurred regularly at short intervals; on 24 February 1556 the princess was replying to Zárate's letters of the 3rd, 9th and 14th of the same month. And it took just a matter of days for the refined silver, sent from the mine a few leagues by land and then along the Guadalquivir, to reach Seville where it was coined. What a contrast with the silver which arrived in Seville from the Indies! Those shipments in the armed galleons of the *flotas* crossed the Atlantic in widely-varying and unpredictable times; the return to Spain, slower than the outward voyage, ranged from 70 to 300 days. Viceroys replied to the king's instructions or queries on mining affairs written months before, and the reply, delayed by the normal wintering of the *flota* in the Caribbean or Gulf of Mexico, would not reach the king's secretary for many months more.

Yet the Indies silver, unlike that from Guadalcanal, never ran out and the annual treasure fleets supplied continuous royal income which served as security for international loans by Genoese financiers. During Philip II's reign Guadalcanal and lesser mines in the Peninsula gave the crown silver worth under one million *maravedís*; on the most recent estimates the crown's bullion receipts from the Indies, both silver and gold, in 1555–1600 amounted to 23 827 million *maravedís*.[63] And this enormous wealth is thought to represent about one quarter of the king's income; it was exceeded by revenue derived from Philip's sharply increased taxation of the Church. The extent of the crown's inherited debts and huge military expenditure are apparent from the inadequacy of income even on this scale; at Philip's death in 1598 the crown owed its creditors 31 875 million *maravedís*.

Gold had been the principal treasure from the Indies and it remained important in the early years of Philip II's reign; it was not exceeded in value by silver until after 1560. Then came the years of the silver boom in the decades after the discovery, during Philip's first regency, of Zacatecas (New Spain; 1546) and the *cerro* of Potosí, the great silver mountain of Peru (1545). Other important strikes were made in New Spain later in the reign. And expectations were heightened by the legend of Inca wealth. In the 1530s Pizarro's band of *conquistadores* had been astounded by the gold and silver objects in the Inca temple of Cuzco; and during the conquest of Peru silver seemed so plentiful that, for lack of iron, Pizarro had his horses shod with the precious metal.[64] Fifty years later Philip

II responded eagerly to the offer of Maria Guarcaycoya, daughter of Manco, king of the Incas, to show mines of silver and rivers of gold, the secret source of her ancestors' great wealth, in the wild forested region of Vilcabamba, north of Cuzco. Philip gave instructions for her to be treated well because of her gesture of good will and out of respect to 'a daughter of that Inca'.[65]

The importance of fostering mining in the Indies was made clear to newly appointed viceroys in the instructions given to them on the eve of their departure from Spain; it was to be one of the priorities of their office. The problems which they found after their arrival were daunting. The terrain presented formidable obstacles. The chief sites were hundreds of miles from the coast and located in arid, mountainous regions. Zacatecas was 8000 feet above sea-level and short of water. Potosí, one of the highest towns in the world, stood in the Andes 13 000 feet above sea level, and to mine the silver mountain which rose behind the town, workers had to ascend to an altitude of 15 000 feet. So intense was the cold in August 1557 that Indians sent to find food froze to death, and Spaniards were killed as their houses collapsed through the weight of snow. According to the historian of Potosí, ten days after this 'two rich seams of silver were discovered and the inhabitants forgot their recent troubles'.[66] The inhabitants were as many as 120 000 in 1570, making Potosí the largest town in Spain or her empire. The lure of silver overcame the discomfort of this most inhospitable location. The remote mining sites also had to contend with wild Indian tribes. In northern New Spain attacks by the Chichimecas cut off Zacatecas in 1561 and repeatedly prevented mining at San Luis Potosí. In Peru in the 1570s the hostile Chiriguanos threatened Potosí; the operators of ore-crushing mills, especially exposed to the attacks because they were in the outlying area, had to be supplied with arquebuses.[67] And further defensive measures were needed to protect Potosí's refined silver which was carried on the backs of llamas to the port of Arica, 400 miles away, the first stage of the long haul to Spain. Fears of attacks there by English corsairs caused the silver to be temporarily stored three leagues inland until the time of loading it on board ship; and in the 1590s the viceroy brought artillery to Arica, and 200 infantry and cavalry under the command of Alonso García Ramón, a 'distinguished soldier' with experience of campaigns in Flanders.[68]

The process used for extracting the silver from its ore in New Spain and Peru was at first smelting, as at Guadalcanal. At Potosí this was done by the Indians who bought ore from mine-owners and smelted it in the open air in their peculiar kilns, perforated to receive the wind and sustain the combustion of the fuel. The kilns were called *guairas*, from the Quichua word for wind. They had their limitations; in 1561 an official informed the king that 'because there is not usually much wind' attempts were being made to replace the *guairas* by other furnaces with bellows to provide the draught, but without success 'either because they do not know how to do it, or because the ore is not appropriate for this'.[69] This shows the relative backwardness of metallurgy at Potosí, and experts from New Spain, where smelting with bellows was established, continued to be sought in the late 1570s.[70]

But a much more important technological change, transforming silver production in the Indies, came with the successful introduction of the cold process of amalgamation,[71] first in New Spain from 1555 and much later in Potosí from 1572. Brief details of how to extract silver 'without fire' by treating ore with mercury had been published in Birunguccio's manual of metallurgy, the *Pirotechnia* (Venice, 1540), where it was described as a profitable technique which did not work with all types of silver ore. In Spain information on the technique was given by a German to Bartolomé de Medina, a resident of Seville – almost nothing is known about him – and as a result he was soon preparing to leave for New Spain to try the method there, hoping to eliminate the high cost of fuel and much of the labour associated with the smelting process, a service to both the operators and the royal treasury.[72] By the end of 1554 after experiments at the silver mines of Pachuca he had found how to apply the amalgamation process to the local ore. Within a year 126 operators were using the new process, and in July 1560 the viceroy, Luis de Velasco, granted Medina sole rights to the method.

The ore was first crushed in stamp mills, powered in New Spain usually by mules because of the scarcity of water; the rotating wheels and gears of the mill caused iron tilt-hammers to fall on the ore. After sifting, the ground ore was piled into heaps in a paved yard where it was then mixed with water, common salt and mercury – about ten pounds for each *quintal* of ore. Gradually the mercury combined with the silver to form silver amalgam, a process which took anything from three to eight weeks to complete, depending on the composition of the very varied silver ore and the environmental temperature. The amalgam was washed in vats, squeezed through linen to remove excess mercury, and finally decomposed by heating – the only stage requiring any fuel – to give high-grade silver, which was melted into bars; the mercury vaporised and was collected after it condensed.

The saving on fuel costs was evident, and the amalgamation process came into its own with ores of low silver content for which smelting was uneconomic. But the savings were offset by the new costs of chemicals needed for Medina's process: common salt and mercury. And some silver ores (sulph-arsenides and antimonides) were difficult to amalgamate. When the crown consulted Hans Schedler, one of the German mining experts in the Peninsula, about silver mining in the Indies he gave sound advice on which of the two methods of extraction was to be used. Neither method would suit all circumstances and the process adopted should fit the local ore; but where wood and charcoal fuel were abundant and cheap, smelting was preferable to amalgamation because it extracted more of the silver.[73] Smelting was not to be eliminated by Medina's new process. At the end of the reign an inspection commissioned by the crown of the mines of New Spain revealed that the amalgamation process was being worked by 296 operators at Zacatecas and nine other mining areas; at San Luis Potosí and three other sites smelting alone was being used by 64 operators.[74]

Amalgamation came slower to Peru; attempts to introduce it in the 1560s had failed. But then, as is well known, in the next decade, through the encouragement of Francisco de Toledo, a viceroy dedicated to mining, processing with mercury became established at Potosí with remarkable consequences for silver production. Pedro Fernández de Velasco had come from New Spain, confident that this could be achieved. In 1571 the viceroy asked him to do tests on Potosí ore in his presence, and when these succeeded further tests were done at Potosí. Fernández was rewarded and paid an annual salary to teach the process to Indians, who in turn instructed others in schools set up in each parish of the town. This technological change was seen by the residents as nothing less than the salvation of Potosí because by this time the richer surface ores had been exhausted and the ores extracted from greater depths were of such low grades that they had become uneconomic to smelt; the operators were moving out and the town seemed to be on the verge of ruin. Now the poorer ores and the abundant residues of former smelting operations became sources of profit and the town enjoyed the peak years of its prosperity. This was also assisted by impressive civil engineering works which guaranteed water supplies to power the operators' numerous stamping-mills for crushing silver ore. Viceroy Toledo was again instrumental. He convened a meeting of mining operators and persuaded the wealthiest to invest in the construction of large reservoirs in the surrounding area, so that the months of drought and slight annual rainfall would no longer be an obstacle in the way of continuous production. In 1573–1621, 32 earth and masonry dams were constructed; an intricate aqueduct system conveyed water from one reservoir to another, and to a central canal, which passed through the town and from which winding masonry canals extended to the sites of the mills; ruined remains of this hydraulic engineering work have survived[75]. Plate 9 shows the process in use.

Although accurate figures of silver production at Potosí can never be recovered because of the unrecorded clandestine traffic in the precious metal, the trends are clear. The effects of the amalgamation process are unmistakeable, and historians of American mining recognise that it caused a revolution in production there, output increasing over seven-fold in 1572–92, and soaring to a peak of nearly 900 000 marks in 1592.[76] The effects of amalgamation in New Spain are not discernible because accounts have not survived for the 1550s, the years immediately following the introduction of the process there.[77]

The silver mines of the Indies were generally assumed to belong to the Spanish monarch; Las Casas, the defender of Indian rights, was a lone opponent arguing that by natural law the mines belonged to the Indians. But unlike Guadalcanal royal administration of the numerous silver mines of New Spain was not attempted; that would have been an enormous undertaking. At Potosí production was left almost entirely in the hands of private individuals.[78] The crown was content with taking the duty: one tenth in New Spain, and the *quinto*, one fifth, in Potosí. It was these royal fractions of the bullion which the *flotas* brought to Spain, along with the more

Plate 9 The great Andean mine of Potosí, chief source of the king's silver. Llamas approach the mine and return laden with ore. In the foreground various stages of the amalgamation process are represented: a large water-wheel raises hammers to crush the ore; in the corral to the left are the *cajones*, characteristic trough compartments in which loads of crushed ore (50 *quintales* at a time in each compartment) were mixed with mercury and salt; in the centre of the corral the resulting silver amalgam is being washed in a large tub to remove impurities prior to thermal decomposition in a furnace; outside of the corral, right of centre, a row of small kilns in which residues from the washing stage were heated. (From a late sixteenth-century MS., Hispanic Society of America MS. K3).

valuable consignments shipped to the mother country by private individuals.

The amalgamation process created a new heavy demand for mercury, and the crown was quick to introduce rigorous controls on its supply not just in order to secure the maximum possible duty from silver production, but also to receive income from selling mercury to the operators. When the princess regent learned from the viceroy of New Spain that mercury would sell there for double the price paid in the Peninsula, she issued instructions on 4 March

1559 in the king's name for all mercury produced in Castile's mine at Almadén, then under royal administration, to be sent to Vera Cruz where the viceroy was to ensure that it was sold 'at the highest price you can get, making the greatest profit after discounting freight and other expenses'. No other mercury could be imported. And on the same day she instructed the officials at the *Casa de la Contratación* to make the necessary arrangements so that ships about to sail for the Indies were carrying 'as much mercury as possible' to satisfy demand in New Spain.[79] Throughout Philip II's reign Almadén would continue to be the principal source of mercury for the mines of New Spain. Potosí on the other hand was supplied from an American source, the rich cinnabar mine of Huancavélica discovered in Peru in 1563. The crown seized possession, declaring null and void all registrations made there, expelling the proprietors and claiming a royal monopoly on this and every other mercury mine in Peru. As at Guadalcanal this led to a law suit against the king, even more bitter and protracted – it lasted over 20 years – and it was secured only after high sums were paid in compensation.[80] There were also reports of armed resistance and threats to kill the viceroy or his agent sent to establish the incorporation of the mines; arrests were made of two mine operators, and a soldier involved in the conspiracy was executed.[81] There are also indications that the *audiencia*, the judicial tribunal, in Lima was lending its support to the mine owners against the viceroy's implementation of the act of incorporation; the crown's letter which instructed the justices of the *audiencia* not to interfere is interesting for associating the incorporation of Huancavélica with the need for funding military campaigns, specifically for Spain's share of the costs of the Holy League against the Turks (1571–3).[82] The arrangements eventually made at Huancavélica – incorporation was not achieved until 1573 – were for the former owners to operate the mines, supplying the crown with stipulated quantities of mercury at a specified price, a contract which in some years they were reluctant to accept; persuasion was achieved by imprisoning the recalcitrant. By 1580 the crown was conceding that, just as in Spain, incorporation in Peru had stifled initiative; any mercury mine discovered would at once be taken for the crown and this removed the incentive to prospect. Philip's response to this was to consider revoking the crown's monopoly on mercury for a period of six years, and he sent instructions to this effect to Martín Enríquez de Almansa, the newly-appointed viceroy of Peru. But in a covering letter, which weakened these orders, the king asked Enríquez not to lift the monopoly until he had discussed the proposal with Francisco de Toledo, the outgoing viceroy, the *corregidor* of Huamanga, and the *audiencias*; if they saw objections the monopoly was to stand.[83] No change was made.

Huancavélica was no light undertaking for the crown. Situated at an altitude of 3800 metres, its climate was harsh and there was no wheat or maize within 20 leagues. Firewood was scarce and fuel supplies, essential for roasting the mercury ore, were eased only with the discovery in 1573 that *ichu*, a local grass resembling *esparto*, would serve.[84] The viceroy gave orders for its conservation, but these seem to have been ineffective; in 1583 shortages caused the furnaces at Huancavélica to be inoperative and Indians had to be sent in search of *ichu* leagues away; the shortages were attributed in part to the consumption of the grass by the pasturing of llamas.[85] Above all there was the extraordinarily difficult transport of mercury to Potosí, 1600 kilometres to the south-east over rough mountainous terrain; there were no roads for wheeled vehicles and the mercury was carried in skin bags on the backs of llamas and mules throughout the 1570s. It was then superseded by a land and sea route using the ports of Chincha and Arica. From the late 1570s the crown freed itself from the problems of transport by contracting others to arrange it. Hopes were raised at the end of the reign by reports of a mercury mine much closer to Potosí, just 20 leagues away. There was the prospect of considerable savings, but nothing came of it, and the more distant Huancavélica remained the source of supply.[86] Production soared from 1830 *quintales* in 1574 to 13 611 *quintales* in 1582, but there were sharp fluctuations.[87] Potosí consumed an annual average of 5000–6000 *quintales* (map 2).

'Mercury is the worst commodity in the world.' The viceroy of Peru who expressed this opinion in a letter to the king explained why this was so: mercury was dangerous and difficult to store, leaked during transport, and had to be given on credit to those engaged in the processing of silver ore.[88] The complications of credit were the consequence of the crown's profiteering; it had made mercury so expensive that the owners of amalgamation plant could not afford it. The mercury which was shipped from Almadén cost the king around 30 ducats a *quintal* and was sold in New Spain for prices varying over the years from 120 to as much as 283 ducats, an average profit of around 100 ducats a *quintal* after deducting freight and administration costs. The profits were less at Potosí; the crown paid around 50 ducats a *quintal* for mercury from Huancavélica, selling it at 75–100 ducats to mercury factors who took it to Potosí selling there at 100–120 ducats.[89]

Map 2 The viceroyalties of New Spain and Peru showing mining centres and other places referred to in the text.

Throughout the reign the operators in New Spain petitioned the king to lower the price of mercury. They even tried to persuade the crown of the greater advantages of importing cheap supplies from China which they alleged could be bought for as little as 7 ducats a *quintal* and shipped to New Spain, selling there at 24 ducats. They said the king's loss of mercury profits would be more than offset by the increased duty received from the boost given to silver processing; and commercial contact with China would allow intelligence to be gathered about that land, so facilitating its conquest and the submission of the Chinese to the royal yoke.[90] That was not undertaken and the crown continued with its policy of assisting operators to buy the mercury through a credit system. When mercury arrived in New Spain it was distributed to the various mining centres and put in the custody of justices, the *alcaldes mayores*. They then allocated mercury to the operators, giving them one year to pay off the value of the quantities supplied, the repayments being collected from the refined silver when this was brought to the royal officials for assay, deduction of royal duty and stamping.[91] But the operators, short of cash for their everyday requirements, instead of using the mercury began to sell it clandestinely at a lower price to others who were reselling it at a profit and at the expense of the crown. And some of the *alcaldes mayores*, paid to administer the credit system, were running their own loan business; at Pachuca and Guanajuato they were found to be providing operators with short-term loans on interest whose collection was given priority over repayment of the crown's mercury loans; the loss to the royal treasury was estimated at 200 000 ducats.[92] The fraud and embezzlement which occurred in this credit system are well known, but the considerable difficulties experienced by the crown in collecting the debts have not been recognised. The visitors who were commissioned to inspect the mines of New Spain in 1597 found that operators still owed the crown over 600 000 ducats for 5000 *quintales* of mercury supplied seven years before, and for a previous consignment given as long ago as 1582.[74] The visitors' comment that the debt was not a great risk because of the securities of the operators' equipment was not very reassuring. As a viceroy of Peru told the king: 'putting operators in prison for the debts or confiscating their engines and equipment would only halt production' and that 'therefore it is impossible to avoid the debts which will always be owing to Your Majesty'.[93] The king's mercury business had failed to bring the desired profits, and had even incurred losses.

Mercury and silver had become the twin foci of the crown's mining policy in the Indies. Viceroy Toledo described Potosí and Huancavélica as the two axles on which the wheels of the entire kingdom of Peru revolved. The crown had to maintain coordinated organisation over great distances to prevent shortages of mercury which would disrupt silver production. The king kept a close watch on reported stocks at Potosí, asking viceroys to ensure that mercury factors bringing it from Huancavélica complied fully with their obligations, and to introduce greater vigilance at Chincha where thefts had occurred during the loading of mercury on ships. New Spain was deprived of a part of its consignment from Almadén in 1565 when a ship caught fire as it was about to sail from Spain, losing 150 *quintales* of its cargo of mercury. And when further losses through the capture or sinking of ships by corsairs were reported, Philip ordered surpluses at Huancavélica to be sent to New Spain to remedy the shortages, a method of provision until then prohibited in order to guarantee supplies at Potosí. Coordination was also needed to phase the six weeks' journey of Potosí's silver by land and sea, and across the isthmus, so that it arrived in time at Nombre de Dios for loading on the fleet; the departing *flota* could not delay because of the limited months of favourable sailing conditions; sometimes the failure of this rendezvous was unavoidable because in some years silver production at Potosí was delayed by the lateness of the rains, preventing the operation of the water-powered mills which crushed the ore.

When Hans Schedler, then directing the Peninsular mercury mine at Almadén for the Fuggers, was asked to advise the crown on the best way to conduct mining in Peru, he replied that the king needed just three things to reap the richest of silver harvests: good organisation, art and industry. The organisation should be provided by sending out 500–600 Germans with administrative experience of mining. The art and industry would be best supplied by drawing on the 15 000–20 000 miners of the Tyrol, though for that the king would have to use his influence with his cousin, the Archduke Ferdinand, who had imposed the death penalty for miners leaving the County; failing that, the Indians should be used because successful mining depended not on theory but on 'habitual practice', and he thought the Indians would be well suited for the task.[73]

The export to the Indies of such large numbers of Germans was never attempted. But at about this time the crown did proceed with another ambitious project involving Ventura Espino, who operated mines in Huancavélica and Potosí. In August 1580 drafts were being prepared in Madrid for a crown contract with Espino to rationalise mining operations in Peru.[94] He would direct production of all the crown's mercury mines combining this with operations at the several silver mines which he owned in Potosí. Espino had to pay all expenses including labour costs and give the king 40 per cent of the profits, keeping the rest. He could recruit masters in mining crafts from Guadalcanal and Almadén – a partial listing of 27 shows these to be all Spanish – and also take Wilhelm Engelbert from the German team operating at Almadén. The king agreed to make arrangements for supplying Indian labour and negro slaves, 400 at first and then a further 100 every year, from Brazil and Cape Verde. Everything was done through the Council of the Indies, from the vetting of the selected craftsmen to the signing of the contract. Unfortunately the contract was not well received in the Indies. When Espino arrived with his recruits they were obstructed by justices; the reasons, not evident, were probably to do with local mining interests. Frustrated and vexed by legal proceedings against him Espino soon died without achieving anything; a dispatch from Lima reported that the project left '60 men in ruin; they had left their land for this kingdom, putting their trust in Your Majesty's word and they now find themselves 2500 leagues from home, deceived and with all their wealth spent'.[95] The emigrant craftsmen had been the victims of the crown's incomplete control of the Indies, impossible to secure from the distance of Madrid. Of more use was the work done by Alonso Muñoz, a mining operator sent by the king to assist Viceroy Toledo's survey of the mines of Peru. Muñoz visited many mines there in the early 1570s, assaying ore, reporting on resources of wood and water, and indicating the availability of local Indian labour. With his help the viceroy was able to compile a register of all mines discovered in Peru. He sent this to the king promising further dispatches of detailed data to keep the Council of the Indies fully informed.[96]

Toledo regretted that Spaniards in Peru were much more

interested in becoming soldiers than proprietors of mines. Still less were they willing to become miners. López de Zúñiga, whom Toledo was to succeed as viceroy, reported that they 'would rather die of hunger than lift a spade'. He advised against using imported negro miners, not just because they would not survive the chill of the Andes, but also because they were unruly, and so many of them would be needed that they would far outnumber the whites, making Peru impossible to govern. Therefore the Indians must work in the mines, although 'it must not be supposed that they will go of their own free will because they are by nature idle'; it was 'right that they should be made to serve and earn a living for themselves, their wives and their children'. The king would be well advised to lift the prohibition on forced Indian labour in mines, because 'if there is no mining there is no Peru'. 'Not much force' should be used, good treatment and pay given, and the transfer of Indians from regions of different climate avoided[97] (the viceroyalty of Peru had some of the sharpest climatic variations on earth).

The dilemma was all too familiar to the Spanish monarchy. Since the first conquests in the New World the crown had struggled to reconcile the needs of the settlers and its obligations to the aborigines. The *conquistadores* had been rewarded with Indian tribute labour, which included work in mines; Ferdinand and Isabella had sanctioned forced labour with pay to extract gold in Hispaniola. Yet the crown viewed the Indians as vassals under its protection; from the Catholic Monarchs to the eighteenth century, the crown was mindful of its duty to civilise and convert the Indians, and to protect them from extreme forms of exploitation. Charles V, influenced by the pleas of Las Casas, tried more than once to abolish the tribute system. At the beginning of his reign he had stopped short of this after receiving Cortes' response that abolition would mean the end of the colonies, because without Indian labour the settlers would be unable to support themselves. Abolition was again attempted with more determination 30 years later by Charles' New Laws (1541–3) for the government of the Indies and good treatment of the Indians. This time the strong feelings of the settlers against change were manifest in the civil war in Peru; Blasco Núñez Vela, the first viceroy, sent out to enforce the New Laws, was killed in the ensuing rebellion. The insurrection continued into the early 1550s. In New Spain opposition also prevented the implementation of the New Laws. Charles had been able to do little more than secure the acceptance of the prohibition on Indian slavery; but forced labour would continue, all the more in demand with the discoveries of the mines of New Spain and Peru.

In the days when mining often meant no more than panning for gold in the river-beds of the Antilles (though there was more arduous excavation as well), Charles V could try to restrict Indian mining labour to 'sifting, washing or other light work'.[98] The conditions of silver mining were far more demanding and Charles was determined to prevent the use of forced Indian labour for this purpose.[99] Faced with the inherited debts of Castile and soaring military expenditure, and presented with the prospects of Potosí and Huancavélica, Philip showed less open opposition to Indian labour in mines. At the end of 1567 he wrote to his officials in Peru that because it was believed that silver production could be much higher if Indians were allowed to work in the mines, this should be investigated and a report sent to the Council of the Indies. In the meantime Indians could be used, provided they came voluntarily, were given a just wage, worked for hours that were not excessive, were not brought from hot to cold regions (or the reverse), and provision made for their religious indoctrination.[100]

There was some voluntary mining by Indians in New Spain, but in Peru they did not come forward to work the deep pits of Potosí or the notorious mercury mines. Viceroy Toledo therefore asked for compulsion to be introduced: 'There is nothing else here but gold and silver, and without these there would be no Spaniards, no religious, nor conversion of Indians'. Compulsion of the Indians was desirable for good spiritual and temporal government; it would save them from idleness, drink and the vices of the flesh; that was why the Incas had kept them employed on works which were of no use.[101] In 1572 he pressed the king on at least two occasions for approval of compulsion: 'Although from Your Majesty's instructions it is presumed that the Indians are to work in mines, Your Majesty knows it will not be with their free will. Your Majesty neither clarifies this nor orders that mines are to be worked without them. A clear declaration from Your Majesty on this is needed.' He said that in Peru it was evident that nothing short of compulsion would bring Indians to the mines; without it the kingdom would collapse, the royal duty of the fifth part of silver would cease, and the Indians succumb to idleness and destruction.[102] Philip never issued a formal statement of approval; instead his response on this delicate subject to a later viceroy was 'consult expert opinion, arrange things as seems best to you, and inform me of what you do.'[103] What Toledo decided to do is well known. Towards the end of 1572 he introduced the *mita*, a system of forced labour in shifts, a repeat of what the Indians had suffered under the Incas. One seventh of adult Indian males was conscripted from every village and forced to work for a specified number of weeks for low pay on roads, in sugar and textile manufactories, and most of all in the mines. A similar system, the *repartimiento*, began to be introduced in New Spain from 1575. Toledo prepared a detailed set of mining ordinances which regulated the conditions of Indian labour and sought to prevent the excesses of exploitation. At Huancavélica, for example, they were to be protected from the dangerous fumes emitted during the roasting of the mercury ore; the uncovering of the vessels in which this was effected was to be done either by an official or a slave, not by an Indian; any Indian who became ill as a result of contraventions of this order was to receive a sum of money and three months medical treatment at the expense of the mine's operator. The hours of work were from one hour and a half after sunrise to sunset with an hour's break at midday for rest and food. Detailed specifications were also given for the materials and dimensions of staircases in pits, with rails in dark shafts for the greater safety of the Indians who climbed up carrying loads of ore.[104]

The king, who had done nothing to prevent this official imposition of forced labour, reacted to some of its effects. When he learned that drainage work at Huancavélica was arduous and caused Indians exhaustion and illness, he gave orders for them to be relieved of it and to be replaced 'by negroes or other types of people'.[105] The alleviation of the Indian miners' burden was also conspicuous in the king's instructions to Alvaro Manrique de Zúñiga on the eve of his departure for New Spain where he was to serve as the new viceroy. Philip had received a disturbing report from the archbishop of Mexico, the outgoing interim viceroy, that work in the mines was the chief cause of the decline in Indian population. The archbishop, who acknowledged that mining must not be allowed to come to a stop, consulted various persons, including the proprietors of mines, to see how working conditions might be improved. They told of the long distances – up to 20 leagues – Indians were taken from their homes to serve in the mines, bringing with them food for 'ten or more days' which went bad; of the lack of housing for Indians who had no proper place to rest or shelter. They recommended homes for Indian families close to the site to avoid their frequent long journeys and encourage them to extend their stay for the income received, thereby facilitating more effective religious indoctrination, since over a longer stay the resident clergymen would have greater opportunity of becoming acquainted with the workers. All of this the new viceroy was asked to investigate in the interests of the good treatment and conservation of the Indians.[106] Missing from this list of proposals was any mention of working conditions within the mines. They were raised by the king in a cool letter to Manrique de Zúñiga, two years after his appointment, chiding him for not doing all he could for the welfare of the Indians.[107] The rebuke was occasioned by reports of abuses: of Indians forced to carry loads of ore to the furnaces; and of a lawsuit in which the king's *fiscal* – a crown lawyer whose duties in the Indies included the protection of the natives – was bringing action against proprietors who had forced Indians to enter a smoke-filled mine, causing the asphyxiation of three of them.

But cases of uncontrolled exploitation of Indians continued to add to the everyday distress of forced labour in abysmal conditions. At Potosí in the 1580s, a concerned proprietor of mines recorded that the hospital was full of sick and injured Indians, 'over 50 of them die each year', and in the tribunals 70 cases relating to Indian deaths were pending.[108] Viceroy Luis de Velasco was shocked by the conditions at Potosí and especially at Huancavélica, where opencast mining had been replaced by work in subterranean galleries. He advised the king that the Indians' burden was intolerable; they were being treated like animals. And he told of the fine dust breathed by Indians as they dug mercury ore, causing a dry cough, fever and eventual death, 'because the physicians regard it as incurable'.[109] The conscience of the king's ministers had been affected, but so long as the king needed silver there seemed to be no escape from the evils of forced Indian labour. The religious were turned to, but some of their replies brought little comfort. The Jesuits of Lima not only condemned the greed of mine operators which led to contraventions of the regulations, but also the regulations themselves; compulsory work in mines was a form of hard labour sanctioned by canon law only for the greatest crimes; in the Roman empire innocent subjects had not been subjected to it. And the Jesuits raised a telling question: if the silver was needed to pay for the defence of the Catholic faith, why should the burden fall so much more heavily on the Indians than on the Spaniards in Peru?[110]

Technological means of alleviating Indian labour were also sought. Innovations were sometimes explicitly presented as contributing to this end, though not only to this end – the goal of optimum production was always an accompanying aim. At Potosí long adits were excavated from the 1580s to provide ventilation and facilitate drainage. At the same time devices for ventilating mines – unfortunately the documents reveal no details and describe them only as 'engines' – were invented by Pedro Cornejo de Estrella of Potosí. They were referred by the Council of the Indies to its *Junta de la Contaduría Mayor*, a committee which had been set up to explore means of increasing colonial revenue. The *Junta* reported favourably and recommended the adoption of Cornejo's invention in Peruvian mines.[111] At Huancavélica excavating adits for ventilation and light proved difficult and Viceroy Velasco, reporting that blasting techniques were being considered, asked for engineers to be sent from Spain 'to do this without risk' because 'no one here knows that art'; his hope was that 'the Indians would work more safely and with more vigour, and consciences would be clear' once the health hazards were removed.[112]

Attention was also given to the improvement of furnaces at Huancavélica. When Pedro de Contreras, who had himself introduced improved designs,[113] gave evidence before royal officials there on the merits of furnaces built by his associate Enrique Garcés, he said they would benefit the health of the natives as well as the treasury.[114] The metallurgy of mercury was simple, requiring no more than the roasting of mercuric sulphide ore, the mercury vaporising and collecting after condensation. But there was scope for making combustion as efficient as possible and reducing contact with poisonous fumes. Garcés' furnaces were reverberatory and designed to save fuel costs. It is not clear if they were ever installed, but the king later gave orders for this to be done for a period of 15 years, after the Council of the Indies had given its approval. Garcés had promised to increase production of mercury by 2000 *quintales* a year, without additional fuel or labour costs; he asked for 4 *pesos ensayados* (1800 *maravedís*) for every *quintal* in excess of existing production. The king wanted a preliminary comparative trial to assess productivity, comparing the weight of mercury produced by 100 labourers using the existing method of roasting, with that produced by the same numbers working with Garcés' furnaces.[115]

Much technological effort was directed to increasing the efficiency of the chemical process of amalgamation. The slowness of this process caused frustration and for decades inventors approached the crown with methods of accelerating it. In New Spain Juan Capellín, a proprietor of mines in Taxco, was awarded a royal licence for a technique which was said to reduce the operation from one of weeks to four days;[116] and in Madrid an Italian promised that amalgamation could be done in just one day.[117] Other proposals to assist the combination of the silver with mercury included the invention of special hammers for crushing ore; an improved sieve; and a method for making pure salt, an essential material for the process.[72]

But the most sought after technique was a way of reducing the quantities of mercury consumed. There seemed to be much wastage; the occurrence of these losses was puzzling[118] and the expense keenly felt. Solutions came in 1587–8, and here the evidence of implementation is firm. In December 1586 news was sent to the king of the discovery in Potosí that the addition of iron scoria to the ingredients of the amalgamation process avoided all loss of mercury and caused a rapid extraction of the silver. The excited royal official who communicated this news, a lawyer of the region's *audiencia*, rejoiced that this invention which so promoted silver production had provided the king's arm with 'the sword of justice to castigate the perfidious English nation'.[119]

But neither the king nor the viceroy was pleased. The king was in the mercury business; while the owners of amalgamation plant would be saved heavy expenses on mercury purchases, the crown stood to lose large sums from the fall in consumption. And the use of the new iron process was spreading, especially after Carlos Corzo[120] had introduced the iron in the readily available form of filings. The viceroy, Fernando de Torres y Portugal, revealed his irritation to the king that 'without either the royal officials or anyone else there telling me, which they should have done, they have already begun to use it'.[121] He decided that immediate action was necessary to preserve the crown's mercury income. In January 1588 he therefore instructed the *corregidor* of Potosí to prohibit Corzo's technique and any other modification of the existing amalgamation process until it was clear whether or not they were 'contrary to His Majesty's service and the public good'; contraventions of the order would be penalised by a fine of 1000 *pesos* (1200 ducats). He had been persuaded by the arguments of the mercury factors. They had said that Corzo's invention was appropriate for lands without mercury, but Peru had all the mercury it needed for the processing of silver. They complained that it was not just to deprive the many who were engaged in its extraction and supply; Huancavélica would be ruined and silver production at Potosí disrupted for two years while necessary changes in equipment were made to work Corzo's process, all of which would greatly reduce the king's income.[122] The king's first reactions were the same, and in 1590 he sent further instructions for the 'cessation of these inventions', urging the viceroy 'to give much attention to this and keep me informed, because nothing could be more important for the processing than the consumption of 7000–10 000 *quintales* of mercury each year which are worth 400 000 *pesos*'.[123]

Neither prohibition took effect. In the *corregidor's* absence his lieutenant suspended the viceroy's order, giving in to the petitions of the local proprietors of mines and mills.[122] And when the *corregidor* returned he endorsed this action, reporting to the viceroy that the proposed ban had 'caused uproar in the town' and that he was reluctant to carry it out because 'it was not in the interests of His Majesty or Your Excellency'. Without Corzo's process the proprietors of the mines and the town as a whole would suffer greatly; it was not just that the interests of the population should be set aside in order to safeguard the few mercury factors. He said their fears were in any case groundless, because the reduction in quantity of mercury needed for each processing operation would be compensated by the overall great increase in the number of operations now that it had become economic to work the poorer silver ores. Huancavélica would therefore continue to prosper, and the king's income would rise not fall.[124]

The mining proprietors of Potosí had certainly suffered in 1587 from the dip in silver production caused by a phase of extraction in which poorer seams were encountered; so any savings in the cost of amalgamation were especially welcome to them at this time. The *corregidor* later saw Corzo's process as a godsend. He said it had restored confidence and encouraged investment; the number of mining operators had increased from 200 to 600, whereas before its introduction 'no more than 25 of 200 mine- and mill-operators were free from imprisonment for debts, possessions were worth less than half their former value and so many fled because of ruin and lack of credit that I was assured this town was about to be depopulated'. And he supported this strong recommendation for the continuation of the iron process with figures taken from the registers of the mercury factorage: in 1586–7, before the introduction of Corzo's technique, 8295 *quintales* of mercury were distributed from the warehouse for silver processing in Potosí; in 1588–9, after its introduction, the quantity rose to 12 146 *quintales*. This beneficial effect was confirmed by an accompanying increase in the receipts of the royal duty on silver by about an eighth.[125]

The king was impressed with the figures and now authorised the adoption of Corzo's method until there were any indications of undesirable consequences.[126] But were the rise in the production of silver and increased mercury consumption due to Corzo's process? There is evidence that in the 1590s Corzo's process was so widely used in Potosí that it created a considerable demand for iron; indeed such was the demand that officials began to ask the crown to order ships sailing for Peru from Seville to be laden with heavy cargoes of iron.[127] This was reminiscent of similar appeals for mercury after the introduction of the amalgamation process. This was not just a request for iron hardware but for its use as a chemical in Corzo's process; that is made clear in another document which reveals that 2500 *quintales* of iron were being consumed annually in the amalgamation process at Potosí, and a further 3000 *quintales* were used to make hammers and other tools.[128] And residents even tried to interest the king in a royal monopoly, bringing the iron from Vizcaya and selling it in Potosí at 25 *pesos* a *quintal*, half the current price.[129] Enrique Garcés, one of the most experienced in Peru's mining affairs, thought the new demand for iron to be sufficiently important to merit a special project. He offered to boost silver production by supplying iron cheaply and ready ground, saving further costs there. The source was the coast of Peru which, he informed the king, was covered with a black sand

and 'Your Majesty needs no more than a lodestone to show that this is pure iron'. Supplies would never run out because 'the continuous waves of the sea batter the rocks which are almost entirely iron'. And he promised that the king's approval of the project would bring lasting profits because 'silver mining in the Indies will never come to an end so long as the world endures'.[130] The outcome of this is not recorded; but the formulation of the project provides clear evidence of the extensive adoption of Corzo's process.

More information about the process comes from an investigation on work at Potosí ordered by the king in 1594. Evidence under oath taken from a mill-owner shows that each load of 50 *quintales* of ground silver ore was mixed with brine, 3 *quintales* of mercury and a suspension of iron filings in water. The quantity of iron added, according to another testimony, was 5–8 pounds. The Indians trod the mixture for five days, and it was heated at night; on the sixth day the combination of mercury and silver was complete. These details show not only that the amalgamation process had been accelerated, but that mercury consumption was reduced to about a half.[131] Meanwhile at Huancavélica surpluses of mercury were accumulating, and in August 1595 the viceroy ordered the work-force to be reduced by one half.[132] The testimony at the inquiry is also interesting for the views given on the function of the iron filings. Its chemical role was not understood; its beneficial action was attributed to the removal of grease and slime from ores, dirt which supposedly consumed the mercury and caused the wastage.[133] From the seventeenth century on it became standard practice to introduce a metallic salt, usually copper, as a reagent of the amalgamation process.

When the silver fleet arrived in Spain after eluding the corsairs lurking in Atlantic waters – no silver cargo was captured during Philip's reign, so effective was the convoy system – the precious metal was consigned in different ways. Some of the bullion belonging to the king was weighed and given to the crown's creditors; some was sent to a mint; much was sold by public auction to merchants who then made arrangements for coining.[134] Ulloa has revealed the considerable royal income from minting after November 1566 when Philip introduced seigniorage on silver coin, perhaps as an alternative to debasing the currency; the king took 50 *maravedís* for each mark of silver coined and 400 *maravedís* for each mark of gold. The receipts in 1595 alone amounted to 200 000 ducats.[135]

The most important mints were in Seville, Toledo and Segovia. To improve the quality of coins and to facilitate their manufacture a notable second mint was erected at Segovia in 1583 with water-powered machinery instead of the usual manual process of manufacture.[136] This was the result of more metallurgical assistance from the Tyrol. The king had secured a small team of craftsmen from the Archduke Ferdinand, and together with parts of the future mint they travelled from Innsbruck to Genoa, Barcelona and Segovia; some of them were to remain for over 30 years. Once the parts were assembled, four water-wheels driven by the river Eresma rotated cylinders, and as silver slabs were passed through these they were compressed to the desired thickness, stamped with the royal arms and cut into rounded shapes. According to Ulloa there are no records of seigniorage receipts from the new mint (the levy was not taken on the coining of silver belonging to the king). The mint which had several floors continued to operate in the eighteenth century.

Assayers were always in demand at mints and mining centres; but they were not easy to find and those appointed were often alleged to be inept. In 1563 the Council of the Indies was considering complaints from the *Casa de la Contratación* that merchants were not buying bullion from the Indies because they had no confidence in the assayers of Peru after experience had shown the ingots to contain less than the required proportion of silver; tests on 200 bars of silver had shown considerable divergence. Whether the result of fraud or ineptness the king agreed that something had to be done; he instructed the viceroy and treasury officials in Lima not to appoint assayers without first making sure that they were sufficiently qualified, and to inform Madrid of all nominees; assayers would in future be subject to fines for their mistakes.[137] And within the Peninsula assayers of mints were indicted in 1585 after coins were found to be short of the legally required quantity of silver. In the ensuing lawsuit which lasted 12 years, the arguments of the defence were accepted that the cause was the variety of techniques used in assaying. As a result the crown appointed an assayer-major to introduce a standard technique and prevent the serious errors of the past.[138]

## *3 A new role for the mine at Almadén*

The processing of silver ore by amalgamation gave new importance to the crown's Peninsular mercury mine at Almadén. Situated in the sierra to the south-west of Ciudad Real it fell within the jurisdiction of the military Order of Calatrava; and when the mastership of the Order was conceded to Charles V, Almadén, along with the rest of its valuable estates and revenues, passed to the crown. Worked since the time of the Romans, Almadén's cinnabar ore continued to be extracted in the early sixteenth century, sometimes under the supervision of royal administrators and for longer periods by lessees.[139] In 1525 Charles leased the mine to the Fugger bankers of Augsburg as compensation for unpaid debts resulting from their heavy loans; it was to be the beginning of their long association with Almadén, bringing German officials and craftsmen to the site.

Before its application to silver production, mercury was in demand, along with its compound corrosive sublimate (mercuric chloride), as medicines, above all in the treatment of syphilis. This medical use would continue; in 1574 Almadén was required by the king to supply 3 *quintales* of mercury every two years for the treatment of sick poor at a hospital in Toledo.[140] Paint was the other principal use of Almadén's ore before the introduction of the amalgamation process; ground up it provided vermilion, the scarlet pigment valued for its intensity and durability. These applications could stimulate production at Almadén to levels which were far from small; in 1547–9 an annual average of 104 *quintales* of vermilion and 1562 *quintales* of mercury were produced, an amount not to be greatly exceeded by the average annual quantity, 1962 *quintales*,[141] sent to the Indies from Seville in 1559–1600 for the amalgamation of silver.

Then came the disastrous fire of 1550. It lasted three months, destroyed equipment and left the mine badly damaged and flooded. The crown took the mine back from the Fuggers and attempted to revive production, but by the end of 1554 this had reached only a quarter of the previous output. A year later the first news of the successful application of the amalgamation process reached the princess regent, and in Brussels Philip began to consider sending mercury from Almadén to New Spain. In April 1557 Ambrosio

Rótulo was appointed royal administrator and governor of the mine. His duties were to restore the mine and produce 'as much ore as possible' sending the mercury extracted to the *Casa de la Contratación* for shipment to the Indies; he was instructed to report developments to the Council of Finance. In 1559 265 *quintales* of mercury, the first of numerous annual consignments from Almadén, was sent from Seville to Vera Cruz. Although Guadalcanal was a much closer silver mining centre it had practically no connection with Almadén, receiving no more than occasional small quantities for testing the practicability of amalgamation. Smelting and not amalgamation remained the established technique of production at Guadalcanal. It was New Spain, where amalgamation predominated, which became the main consumer of Almadén's mercury, and in 1559 Philip declared a royal monopoly on this export trade. The king's determination to enforce it is evident in his strong reaction to reports that merchants in Seville were secretly buying large quantities of mercury, taking it by night on boats along the Guadalquivir for loading on ships to sell in New Spain. After the matter was discussed in the Council of Finance, Philip appointed a justice to conduct an inquiry in Seville and Cádiz to find out who had been involved in the clandestine traffic and 'on whose ships, in what quantities, with whose connivance and by which correspondence with the Indies'.[142]

At Almadén the mercury produced was poured into sheepskin containers, and carried by mule or in carts pulled by oxen 33 leagues to Seville. It was delivered to royal officials who untied the skins and weighed the contents; they then repacked the mercury in skins which were enclosed within sealed barrels and finally in wooden boxes, stamped with the royal arms and loaded on board ships of the *flotas*. When 96 *quintales* of mercury arrived at the *Casa de la Contratación* in August 1562, the officials asked the king to provide suitable facilities because 'it needs great care in handling and storage in a place paved with brick so that if there is any spillage it can be recovered; it is so subtle that it seeps through the containers and much could be lost'.[143]

Production under royal administration remained low. In November 1557 Francisco de Mendoza had been sent from Guadalcanal to inspect the mercury mine. His report revealed that much needed to be done.[144] The pits were deep – over 600 feet deep – and lacked effective drainage. Lack of available fuel supplies had forced the roasting of ore to be done three leagues from the site, the closest source of timber, erecting furnaces in the countryside; Mendoza was worried about the opportunities this gave for theft because there was no check on what was produced there. He inspected another forest but was dismayed to find that it had been largely destroyed by fires lit to provide pasture for sheep. Competition for land and timber was a recurring difficulty at Almadén. When Rótulo imposed prohibitions on cuts of timber in forests traditionally used by the residents of the neighbouring town of Chillón, he received threats of assassination. And in 1568 the king had to declare against farmers who wanted to plough and cultivate land which the mine's administrators wanted to preserve for the growth of forests.

The geographical *relaciones* of 1575 describe Almadén as a town of 400 householders, mostly poor peasants, located in a hot, mountainous region; there was a small river two leagues away which dried up in summer; bread and wine supplies depended on imports from surrounding areas. Shortages of bread would frequently interfere with production at the mine. So did the constant labour shortage; long delays in the payment of wages – up to eight months in 1560 – hindered the recruitment of local labour, and during harvest time men left the mine to gather what little wheat could be cultivated. Work in the mine was hardly inviting. The hardness of the rock made excavation arduous. Hollows had first to be made by hammer and pickaxe; iron wedges were inserted in the hollows and struck with hammers. The heavy stone extracted in this way was then broken into pieces.[145] The unpleasant and unhealthy conditions within the mine were witnessed by the chronicler Ambrosio de Morales. Curious to see the inside of a mine, he had visited it as 'a young lad' and found 'a true representation of hell'. The pumps worked constantly to keep the mine drained; but the miners could 'not bear more than a very few hours each day'. The stench of sulphur was intolerable, and 'everything else there is horrible and dreadful. Those who work there have a constant tremble and generally do not live long because, as Pliny also noted, mercury gradually penetrates to the bones'.[146]

Faced with all the difficulties on the site and the expense of reconstruction, still not completed after the fire, the crown decided to pass the burden to the Fuggers; they would remain lessees until the middle of the seventeenth century with a succession of ten-year contracts. In these the Fuggers aggreed to supply the king with specified quantities of mercury at an agreed price, which was subject to a charge of ten per cent interest if the king delayed payments; the lessees would meet the costs of the fire damage. The king authorised the requisitioning of beasts and carts from anywhere in the Peninsula for the transport of mercury to Seville. That was an important provision because there had been shortages; in a land without navigable waterways they were indispensable. Legal proceedings were later brought by the crown against men who, entrusted with this requisitioning, had taken bribes from owners reluctant to part with their animals and carts, resulting in 'great damage to the treasury'.[147]

The contract of August 1562 required the Fuggers to produce 600 *quintales* in the first year, 800 the next, rising to 1000 in the remaining eight years; selling to the king at 25 ducats a *quintal*. The target was soon achieved and the contract modified to raise annual production up to 2000 *quintales*, the king's cost rising to 29 ducats. And in the 1590s annual production often reached 3000 *quintales*.

How had the German administration succeeded? Two important innovations were introduced during the Fuggers' tenure of lease. First, around 1570, the furnaces were improved, saving labour and reducing the requirements for wood fuel in an area where it was only secured with difficulty. Before this change the ore, placed in covered earthenware jars, 32 to a furnace, had been heated from below; the mercury produced vaporised, and after the furnace had cooled the jars were opened and the condensed mercury collected with a ladle. Now reverberatory furnaces were introduced. Hans Schedler, the director of operations at Almadén, explained the advantages to the king.[148] The flames were now directed towards the sides of the furnace and reflected on to the jars of ore; this technique of applying heat was found to be so much more efficient that up to 180 jars of ore could be roasted with the same quantity of wood as was needed before to heat 32 jars. And in addition a receiver positioned beneath the jars collected condensed mercury over a period of three months instead of the previous chore of spooning it out daily.

More important the Fuggers solved the problem of labour shortage through the use of convicts. Schedler had come to the crown with this request saying that unless 30 slaves or convicts were supplied, the contract could not be fulfilled. In February 1566 the king authorised the allocation of this number of men from Toledo jail who had just been sentenced to serve on galleys; in 1583 this was increased to 40 because of 'deeper pits and more arduous work'. The king stipulated that this was on condition that the Fuggers took care to supply adequate food and clothing and released the convicts as soon as their sentences had been served, because 'although galley labour may be harder than the mine it is not my wish that they be harmed'.[149] Labour was also found in the aftermath of the Granada uprising. During the crown's subsequent transport of Moriscos, distributing and resettling them in various parts of Castile, Schedler was granted 93 of them to work in the mine as wage labourers; the measures of Don Juan of Austria, the king's military commander who put down the uprising, for avoiding large concentrations of Moriscos were waived for Almadén. And in 1588 authorisation was given for an additional supply after 200 Morisco families were found to be living in Ubeda, Baeza and Jaén, illegally approaching their homeland; the death penalty prescribed for this offence was commuted to settlement at Almadén, but surprisingly without forced labour – voluntary work for pay was expected instead.[150]

Concerned over the welfare of the convicts, the crown ordered an investigation in 1593. Were the convicts receiving adequate food and medical care, working moderate hours, and given release when their sentences had elapsed? Had death or illness resulted from the failure to observe these terms of the contract? Evidence was taken in secret from 12 convicts (thieves and brigands, but also a friar convicted of murdering his lover's husband), overseers, masters of furnaces and the barber-surgeon of the mine's infirmary.[151] The food and clothing were found to be adequate, and convicts had been released on the completion of their sentences. The medical treatment was generally praised, though some convicts complained that insufficient time had been allowed for convalescence. But there was disturbing evidence of overwork in the recent past in the interests of greater production. There were allegations of cruel taskmasters, beating exhausted workers and forcing some to take roasted ore from jars still hot, causing burns and the inhalation of dangerous fumes. None of those who gave evidence knew of the existence of the king's instructions, which limited the hours of work and contained other provisions for the care of the convicts.

Most of the convicts did not sign their testimony because they did not know how to write; two master craftsmen, not convicts, failed to sign because their trembling hands, a symptom of mercury-poisoning, prevented it. The outcome of these proceedings is not known.

The king's concern for the convict miners of Almadén was in harmony with his attention to the welfare of Indians in the mines of the New World. Was this part of an overall policy to promote the health of his subjects?

## *Notes*

1. T. González, *Noticia histórica documentada de los célebres minas de Guadalcanal desde su descubrimiento en el año de 1555, hasta que dejaron de labrarse por cuenta de la real hacienda*, 2 vols., (Madrid, 1831), vol. 1, p. 204; the comment came from García Martín de la Bastida, who had acquired a share in the mine, in a letter to the royal secretary Antonio de Eraso, 15 June 1556. This work by the nineteenth-century archivist Tomáso González is an important collection of hundreds of documents relating to Guadalcanal, assembled during his official appointment to restore the Archivo General de Simancas after the damage caused by Napoleon's soldiers who had lodged there. Although this collection of primary sources is quite well known, strangely it has not been used to write a history of the mine's development. González's motive was in part to stimulate a revival of interest in these long-abandoned mines at a time of silver shortage.
2. Ambrosio de Morales, *Las antigüedades de las ciudades de España* (Alcalá de Henares, 1575).
3. Jerónimo Muñoz, *Libro del nuevo cometa* (Valencia, 1573), p. 30v. Unfortunately Muñoz was unaware that the mine was discovered several months before the appearance of the comet.
4. González, *op.cit.*, vol. 1, pp. 378–82. Similarly in April 1573 Agustín de Sotomayer, who had come to Spain from Mexico, promised the king that further discoveries of silver mines must occur because the planets were constantly generating the metal in Spain; a crown contract for exploration close to Guadalcanal was agreed with him later that year; *ibid.*, vol. 2, pp. 405–8 and 421–8.
5. *Ibid.*, vol. 1, p. 53f.
6. *Ibid.*, *ibid.*, p. 60f.
7. *Ibid.*, *ibid.*, pp. 78f, 89f and 112f.
8. *Ibid.*, *ibid.*, p. 96f.
9. *Ibid.*, *ibid.*, p. 110.
10. *Ibid.*, *ibid.*, p. 281.
11. *Ibid.*, *ibid.*, pp. 253–7.
12. AGS: PR 26/159.
13. González, *op.cit.*, vol. 1, p. 98f.
14. *Ibid.*, *ibid.*, p. 115; the princess regent to the justices of Almodóvar and other *pueblos*, 11 March 1556.
15. *Ibid.*, *ibid.*, p. 96.
16. *Ibid.*, *ibid.*, p. 117.
17. AGS : E 511/125 the king to Anton Fugger, 9 May 1556, Brussels.
18. González, *op.cit.*, vol. 1, p. 357; the estimate of 1000 came from *licenciado* Murga, *alcalde* at the site, in a request for greater provision for mass, 'not just on Sundays but every day for those who want it'.
19. AGS : PR 26/158.

20. González,*op.cit.*, vol. 1, p. 225.
21. AGS : *Escribanía mayor de rentas: minas, legajo* 14, *número* 6.
22. González, *op.cit.*, vol. 1, pp. 275 and 282–3.
23. AGS : CJH 68/124, Martín López de Yeribar, *contador*, and Gerónimo de Anuncibay, treasurer of the mine, to the king, 4 January 1566, Guadalcanal.
24. González, *op.cit.*, vol. 1, p. 262.
25. AGS : *Escribanía mayor de rentas: minas, legajo* 14, *número* 6.
26. González, *op.cit.*, vol. 1, p. 469.
27. AGS : CJH 103/12, Martín López de Yeribar and Gerónimo de Anuncibay to the king, 8 January 1570, Guadalcanal.
28. González, *op.cit.*, vol. 1, pp. 129–30.
29. AGS : DC 8/54, the king to the governor of the province of León of the Order of Santiago, November 1579, Madrid.
30. P. Bakewell, 'Registered Silver Production in the Potosí District 1550–1735', *Jahrbuch für Geschichte von Staat, Wirtschaft und Gesellschaft Lateinamerikas,* **12** (1975), 92.
31. Assuming one contemporary official valuation of 2320 *maravedís* for one mark of fine silver.
32. E. Lorenzo Sanz, *Comercio de Espãna con América en la época de Felipe II*, vol. 2 (2nd ed. Valladolid, 1986), p. 252.
33. González, *op.cit.*, vol. 1, p. 489.
34. *Ibid.*, p. 383 f.
35. IVDJ 31 'o', unfoliated, anonymous legal advice for the king.
36. González, *op.cit.*, vol. 1, p. 447.
37. AGS : DC 46/99, Francisco de Almaguer, *contador mayor* of the *Consejo de Hacienda* to the king, February 1564, Madrid, summarising legal advice.
38. *Ibid.*
39. González, *op.cit.*, vol. 2, pp. 223 and 252.
40. This was a royal provision for the Indies, but presumably the king's disapproval also applied to the Peninsula; *Recopilación de leyes de los reynos de las Indias* (Madrid, 1681), *lib.*i, *tít.* xii, *ley* iv, Philip II, 15 November 1592, Viana, Navarre.
41. *Novisima recopilación de las leyes* (Madrid, 1614), *lib.*vi, *tít.* xiii, *ley* iv, Act of Incorporation of mines, 10 January 1559; *lib.* vi, *tít.* xiii, *ley* v, new mining ordinances, 18 March 1563. This organisation was modified by the regulations of 1584, which provided for an unspecified number of local mining administrators under the administrator-general, all appointed by the Council of Finance; these administrators were responsible for enforcing the mining regulations and were given jurisdiction over civil and criminal cases in mining districts; *ibid.*, *lib.* vi, *tít.* xiii, *ley* ix, 2 August 1584. The ordinances do not seem to have had the desired effect; the number of mining claims and registrations in the 1570s has been taken to signify a weakening interest among the king's subjects: M. Ulloa, *La hacienda real de Castilla en el reinado de Felipe II* (Madrid, 1977), p. 464. Apart from lack of money for investment, the most likely reason for the lack of response, one wonders how much opposition was aroused by Philip's reassertion of monarchical rights. There is record of some aristocratic opposition. In 1573 Bernabé Manjon, administrator of mines in the district of Almodóvar, went to the Campo de Calatrava to implement the royal ordinances with plans to activate the unworked mines of the marquis de la Guardia and to prospect in the area. The marquis rejected the king's ordinances which were shown to him and which required him to allow prospecting on his land and the working of dormant mines. Adamant that the mines were his property and not the king's, he went in person to the site where operators and prospectors had begun to arrive and so intimidated them that they fled; AGS : DC 8/61. There is also evidence of official dissatisfaction with the results of the mining ordinances; the king's councillors were then blaming the act of incorporation with its restrictive measures for the slight interest in mining ventures; IVDJ 76/61, Hernando de Vega, president of the *Consejo de Hacienda* to the king, 13 March 1581.

42. F. Ruiz Martín, 'Las finanzas españoles durante el reinado de Felipe II', *Cuadernos de Historia,* **2** (1968), 118–21.
43. González, *op.cit.*, vol. 2, p. 94.
44. *Ibid.*, *ibid.*, pp. 121–7. Mendoza was not party to these economies; he had died a few months before and his office of administrator-general of mines remained unfilled for 20 years. In May 1568 an offer of employment from the *Consejo de las Indias* attracted some interest amongst those once employed at Guadalcanal. They were offered work in the mines of New Granada, around Bogotá; some were willing to emigrate provided the wages were as good as those they had been used to at Guadalcanal, and they were assisted in the costs of shipping their possessions; *ibid., ibid.*, pp. 224–5.
45. AGS : CJH 68/123, Martín López de Yeribar, *contador*, and Gerónimo de Anuncibay, treasurer, to the king, 4 September 1566, Guadalcanal.
46. AGS : CJH 69/66, 67 and 71, residents of Constantina to the king, undated. The extent of the crown's rights in Constantina is unclear. At the time of the princess regent Constantina had been referred to as within the jurisdiction of Seville. But the later documents indicated here speak of an impending 'alienation from the crown'. Had Philip recently acquired the town? It is however certain from these sources that the king received annual income from the *alcabala*, the sales tax levied in the town, and that he was contemplating the sale of the rights to this revenue.
47. González, *op.cit.*, vol. 2, p. 172, royal *cédula*, 11 May 1565.
48. *Ibid.*, *ibid.*, p. 278f.
49. *Ibid.*, *ibid.*, p. 287.
50. *Ibid.*, *ibid.*, p. 312f.
51. *Ibid.*, *ibid.*, pp. 365 and 369.
52. *Ibid.*, *ibid.*, pp. 351 and 357.
53. *Ibid.*, *ibid.*, pp. 386–8.
54. The rise in production over this period is detectable in the figures tabulated by González, *ibid., ibid.*, p. 582. A higher estimate of 300 000 ducats for the value of silver from the new seam was given by the mine's officials, *ibid.*, *ibid.*, p. 458, but this is not supported by the production figures.
55. *Ibid.*, *ibid.*, p. 461f.
56. *Ibid.*, *ibid.*, p. 475f.
57. AGS : DC 46/27, 'Relación y inventario de todas las cosas que hasta agora se an entregado a Pedro Martín del Freyle depositario nombrado dello'. This, the fullest inventory, was made in 1583.
58. IVDJ 76/64, Hernando de Vega to the king, 13 March 1581. The amalgamation process had been tried at Guadalcanal on several occasions before this. Just one month after the start of operations there the princess regent wrote to Zárate that she had heard from New Spain of a quicker and cheaper method of extracting silver using mercury; she wanted to know if the Germans knew how this was done. Soon Mendoza was supervising trials of the technique, but he was 'sceptical that silver could be extracted without fire'. The tests did not succeed and in September 1558 he wrote to the king that the amalgamation process was unsuitable for the ores of Guadalcanal but might be suitable for silver ore elsewhere in the Peninsula. In the early 1560s Mosén Boteller was brought over from New Spain to treat the residues of the smelting process with mercury; this also was unsuccessful. González, *op.cit.*, vol. 1, pp. 79, 416, 490, 555; vol. 2, pp. 87f, 198f, and 379.
59. Ulloa, *op.cit.*, p. 466.
60. T. González, *Registro y relación general de minas de la Corona de Castilla*, vol. 2 (Madrid, 1832), pp. 47–8.
61. These figures from González, *Noticia*, vol. 2, p. 582 are not just for Guadalcanal, by far the most important producer, but also include small yields from Aracena, Cazalla and other lesser mines.

62. AGS : DC 46/99, Francisco de Almaguer, *contador mayor*, to the king, February 1564, Madrid.
63. Sanz, *op.cit.*, vol. 2, p. 253.
64. W. H. Prescott, *History of the Conquest of Peru* (revised edition, London, 1888), p. 213.
65. AGI : L 570, *lib.* 15, f. 27v, the king to García Hurtado de Mendoza, viceroy of Peru, 20 January 1589. Two years later a letter from the viceroy mentioned 'the hoax of Vilcabamba'. Was he referring to Maria's offer? The hoax was certainly concerned with mining and had deeply embarrassed him; so much so that when some ore was later discovered there and operators asked him to provide native workers he was 'reluctant to have anything to do with it'. Eventually he sent 200 Indians there; *Gobernantes del Perú*, ed. R. Levillier, 14 vols. (Madrid, 1921–6), vol. 12, p. 229 and vol. 13, p. 16.
66. B. Arzáns de Orsúa y Vela, *Historia de la Villa Imperial de Potosí*, ed. L. Hanke and G. Mendoza (Providence, R.I., 1965), pp. 102–3. This history was written in 1705–36.
67. *Licenciado* Matienzo,*oidor* of the *audiencia* of Charcas to the king, 23 December 1577, Potosí; *La Audiencia de Charcas. Correspondencia de presidentes y oidores*, ed. R. Levillier, 3 vols. (Madrid, 1918–22), vol. 1, p. 463.
68. *Gobernantes del Perú*, *op.cit.*, vol. 13, p. 131.
69. *Ibid.*, vol. 2, p. 535.
70. *Audiencia de Charcas*, *op.cit.*, vol. 1, pp. 461 and 468.
71. A full account of the process in the Indies with technical detail and the reproduction of numerous documents from the Biblioteca Nacional, Madrid, are given in M. Bargalló, *La amalgamación de los minerales de plata en Hispanoamerica colonial* (Mexico, 1969).
72. L. Muro, 'Bartolomé de Medina, introductor del beneficio de patio en Nueva España', *Historia Mexicana,* **13** (1963–4), 517–31 reproduces Mexican documents revealing Medina's intentions.
73. IVDJ 22c/98a–98d, Hans Schedler to the king, c. 1580.
74. AGI : *sección de Gobierno : México, legajo* 24, *ramo* 1, *número* 7a, 'Relación del estado que tienen las minas desta Nueva España y las de Zacatecas y lo que deven los mineros a su magd.', 24 April 1598, Mexico.
75. Photographs of the remains of mills and masonry canals are included in W. E. Rudolph, 'The Lakes of Potosí', *The Geographical Review*, **26** (1936), 529–54.
76. Bakewell, 'Registered Silver Production in Potosí', *op.cit.*, and 'Technological change in Potosí: The Silver Boom of the 1570s', *ibid.*, **14** (1977), 57–77. Tables of production figures and graphs rising to peaks clearly show the same trends in A. Jara, *Tres ensayos sobre económica minera hispanoamericana* (Santiago de Chile, 1966), pp. 111–16; but the peak year given here is 1586, and Jara comments that the wide fluctuations in production after this year are not readily explicable.
77. Production figures for Zacatecas have been calculated by Bakewell from payments of royal duty; for the years 1560–1600 they show an average annual production of just under 125 000 marks; P. Bakewell, *Silver Mining and Society in Colonial Mexico : Zacatecas (1546–1700)*, (Cambridge, 1971), p. 246.
78. According to Bakewell there were at least two amalgamation plants under royal administration in Potosí in the 1570s; the total number of plants there is said to have been 65 in 1600; P. Bakewell, 'Mining in Colonial Spanish America', *The Cambridge History of Latin America*, ed. L. Bethell, vol. 2 (Cambridge, 1984), pp. 105–51.
79. Diego de Encinas, *Cedulario indiano* (Madrid, 1596), vol. 3, pp. 416–7, royal *cédulas*, 4 March 1559, Valladolid. Two years before this the king wrote from Brussels to the Council of Finance that it seemed more profitable for Almadén mercury to be sent to New Spain than to be sold in the Peninsula; González, *Noticia, op.cit.*, vol. 1, p. 387.
80. G. Lohmann Villena, *Las minas de Huancavélica en los siglos XVI y XVII* (Seville,

1949), pp. 29, 37, 67, 105, 139 and 143 show the crown's difficulties with the persistent Amando de Cabrera.

81. *Gobernantes del Perú, op.cit.*, vol. 7, pp. 149–50, Dr. Loarte to the king, 13 March 1573, Huamanga.
82. AGI : L 1/29, *libro* 10, folios 127–31, copy of a royal provision, 25 January 1574, Los Reyes.
83. *Cedulario indiano, op.cit.*, vol. 3, pp. 425–6, the king to Martín Enríquez, 30 September 1580, Badajoz.
84. Lohmann, *op.cit.*, p. 48f.
85. AGS : GA 145/98, information communicated to the crown by Fray Bernardo de Gongora, 1583.
86. *Audiencia de Charcas, op.cit.*, vol. 3, p. 303, *audiencia* of Charcas to the king, 23 February 1596, La Plata.
87. From figures in Sanz, *op.cit.*, vol. 1, p. 497.
88. *Gobernantes del Perú, op.cit.*, vol. 13, p. 252, García Hurtado de Mendoza to the king, 20 January 1595, Los Reyes.
89. All figures from Sanz, *op.cit.*, vol. 1, p. 495.
90. AGI : P 238 ii/3, recommendations by Bartolomé de Laguila and other mine operators of New Spain, 21 April 1584, Mexico City. The importation of Chinese mercury was later seriously considered. The archbishop of Mexico, a former interim viceroy, recommended it to the king and said he would write to the governor of the Philippines to see how much it would cost to bring Chinese mercury to Acapulco; although 'it would give much money' to the infidel Chinese, it would also 'benefit Your Majesty's subjects'; AGI : P 239/20, archbishop of Mexico to the king, 22 January 1589. And Luis de Velasco, viceroy of Peru, wanted the same because he was appalled by the conditions which the Indians had to endure in the Huancavélica mine. There is no record of Chinese imports.
91. The king's instructions for both New Spain and Potosí were for half of the cost of the mercury to be paid for immediately, giving only one half on credit 'on the shortest possible terms'; *Recopilación de leyes de las Indias, op.cit., lib.* viii, *tít.* xxiii, *ley* x, 18 May 1572 and 26 March 1577. But viceregal orders only referred to credit for a period of one year; AGI : P 182/41, Enríquez de Almansa, viceroy of New Spain, to the *alcalde mayor* of the mines of Taxco, 11 December 1572, Mexico City.
92. AGS : GA 120/296, Hernando Jaramillo to the king, 16 August 1581. To remedy this abuse, apart from introducing severe penalties for offending *alcaldes*, a distinctive royal emblem was stamped on silver destined for repayments of the mercury loan.
93. *Gobernantes del Perú, op.cit.*, vol. 13, p. 167, García Hurtado de Mendoza to the king, 24 April 1594.
94. AGI : P 238i/3 and 4 contain several documents relating to this *asiento*; AGI : P 238i/4, terms of the final contract, 2 December 1580, approved by the king; and a later *consulta* of the Council of Finance.
95. AGI : L 127, Juan de Medina Avellaneda to the king, 1 December 1586, Lima.
96. *Gobernantes del Perú, op.cit.*, vol. 3, p. 576, Francisco de Toledo to the king, 1 March 1572, Cuzco. Muñoz may have been sent from Guadalcanal; the viceroy's letter to the king in June 1570 referred to mining experts sent to him from Guadalcanal; *ibid., ibid.*, p. 430.
97. *Gobernantes del Perú, op.cit.*, vol. 1, p. 529, Diego López de Zúñiga, conde de Nieva, viceroy of Peru, to the Council of the Indies, 31 August 1563, Los Reyes.
98. *Colección de documentos para la historia de la formación social de Hispanoamérica 1493–1810*, ed. R. Konetzke, vol. 1 (Madrid, 1953), pp. 97–8.
99. *Ibid., ibid.*, pp. 282–3.
100. AGI : P 275/73, the king to the viceroy and *audiencia* of Lima, 27 December 1567, Escorial.
101. *Gobernantes del Perú, op.cit.*, vol. 3, p. 430, Francisco de Toledo to the king, 10 June 1570, Los Reyes.

102. *Ibid.*, vol. 4, pp. 385 and 427, Toledo to the king, 24 September 1572, Cuzco and another letter of 1572, reproduced without date.
103. AGI : L 570, *lib.* 15, f. 27, the king to García Hurtado de Mendoza, 20 January 1589. Similarly the viceroy of New Spain was told 'you are to arrange what seems best to you; we trust you will do what is most suitable', *Colección*, ed. R. Konetzke, *op.cit.*, vol. 1, p. 483, the king to Martín Enríquez de Almansa, 7 May 1574, Madrid. And Philip's connivance in forced labour in the silver mines of Honduras and Guatemala, just when he had been instructing Franciscan friars to spy on proprietors of mines there to see if compulsion was being used, is described in W. L. Sherman, *Forced Labour in Sixteenth-century Central America* (Lincoln, Nebraska, 1979), p. 232 f.
104. *Gobernantes del Perú, op.cit.*, vol. 8, pp. 143–240. Francisco de Toledo's mining ordinances, 7 February 1574, La Plata.
105. *Colección de documentos inéditos relativos al descubrimiento, conquista y organización de las antiguas posesiones españolas de América y Oceanía*, ed. J. Pacheco *et al.*, vol. 19 (Madrid, 1873), p. 174; the king to Luis de Velasco, viceroy of Peru, (n.d.). Philip did not consistently encourage the use of negroes for mining in Peru. In 1590 when Indian labour at Potosí was said to have been short because of epidemics of smallpox and measles, the viceroy agreed to the request of proprietors of mines for black slaves to be brought 'from Brazil and other parts'. But Philip advised the viceroy to reconsider this because negroes tended 'to dominate and maltreat the Indians'; another objection was that the proposed slave traffic would encourage evasion of the royal duty by the smuggling of silver along the back-door route from Potosí to Río de la Plata; AGI : L 570, *lib.* 15, ff. 107–8, the king to García Hurtado de Mendoza, 9 October 1591, Escorial; for the viceroy's letter of 30 December 1590, *Gobernantes del Perú, op.cit.*, vol. 12, p. 201 f.
106. *Colección*, ed. Konetzke, *op.cit.*, vol. 1, pp. 558–9, the king to Alvaro Manrique de Zúñiga, marqués de Villamanrique, 25 May 1585, Barcelona.
107. *Ibid.*, *ibid.*, p. 581, same to same, 8 August 1587, Escorial.
108. Luis Capoche, *Relación general del asiento y Villa Imperial de Potosí* [1585], ed. L. Hanke (Madrid, 1959), pp. 158–9.
109. *Gobernantes del Perú, op.cit.*, vol. 14, p. 37, Luis de Velasco to Philip II, 10 April 1597, Lima; and pp. 267–8, Luis de Velasco and the *audiencia* of Lima to Philip III, 30 April 1600, Lima.
110. AGI : L 33, *lib.* 3, ff. 84–6v, reply of the Jesuits of Lima to questions on Indian labour in mines, 16 January 1599, Lima; the Franciscans, on the other hand, did not question the licitness of compulsory labour, but merely recommended the avoidance of excessive exploitation and the exclusion from the mines of Indians engaged in agriculture, since otherwise famine would result; *ibid.*, pp. 74–5v, 12 January 1599, Lima.
111. AGI : L1, *número* 86, *consulta, Junta de la Contaduría Mayor*, 11 February 1588, Madrid.
112. *Gobernantes del Perú, op.cit.*, vol. 14, pp. 267–8, Luis de Velasco and the *audiencia* of Lima to Philip III, 3 April 1600, Lima.
113. His furnaces were used for decades at Huancavélica. They were rectangular and surmounted by a dome; after the fire was lit mercury vapour passed through long tubes and condensed on the walls of the furnace; Lohmann, *op.cit.*, p. 128.
114. AGI : P 239/16, ff. 49v–51, evidence of Pedro de Contreras, 11 January 1582, Oropesa. The investigation was ordered by Viceroy Enríquez. He said that although Garcés' furnaces had been tested in Lima he wanted further trials at Huancavélica; *ibid.*, ff. 40v–41, Enríquez de Almansa to the *corregidor* and officials of the royal treasury at Huamanga, 20 June 1581, Lima.
115. AGI : L 581, *lib.* 11, ff. 1–2, the king to García Hurtado de Mendoza, viceroy of Peru, 17 July 1593, Escorial.
116. Bargalló, *op.cit.*, pp. 150–2.
117. AGI : P 238 ii/2, Francisco Lombardo to the king, 2 March 1584, Madrid.
118. Juan de Cárdenas, a natural philosopher in Mexico, rejected various current

explanations: that mercury, so much resembling silver, was during the amalgamation process converted into silver; that mercury was lost during the washing of the ore; that mercury disappeared by sinking into the earth. He believed instead that the mercury was disappearing in the form of fumes and heat; *Problemas y secretos maravillosos de las Indias* (Mexico, 1591), p. 90f. The mechanism of the amalgamation process, even today, is not easily described; it is complicated by side reactions. The problem of reducing the mercury consumed was solved entirely empirically without any understanding of the reaction mechanism. At the time it was an outstanding problem which attracted widespread attention. From Milan, Sebastiano Grinco, a 'physician and distiller', offered the king his method of producing silver with 'very little loss of quicksilver', AGS:E 1139/199, Sebastiano Grinco to the king, 26 March 1573, Milan.

119. *La Audiencia de Charcas, op.cit.*, vol. 2, p. 259, *licenciado* Cepeda to the king, 9 December 1586, La Plata.
120. Since doubts have been expressed about his origins it might be as well to indicate here that he himself stated that he was the nephew of Juan Antonio Corzo; AGI : L 273, f. 169, Carlos Corzo to García Hurtado de Mendoza, 4 February 1590, Potosí. Juan Antonio, a merchant of Seville and of Corsican origin, became one of the richest men of his age, leaving 1 600 000 ducats at his death in 1597. The above document also clarifies the motive which led Carlos to his iron process; he had been a mercury factor arranging for the carriage from Huancavélica to Potosí when the market collapsed because much more of the silver ore extracted at Potosí had become poorer and uneconomic to process with mercury. With unsold mercury on his hands and deep in debt Corzo had begun urgent research on ways to process poor ores; he discovered his technique around 1587. The use of iron scoria which had preceded it was due to Garci Sánchez; there were other claimants to the invention and the disputes led to lawsuits; *Audiencia de Charcas, op.cit.*, vol. 3, pp. 197–8, Jerónimo de Tovar y Montalvo, *fiscal* of Charcas, to the king, 28 March 1593, La Plata. And in New Spain the royal cosmographer Francisco Domínguez had been working for nine years on the improvement of the amalgamation process with iron; it was said to have been successful at Taxco and Pachuca; AGI : P 261/9, f. 60v, evidence given before the *audiencia real* of New Spain by Bartolomé Palomino, proprietor of mines in Guanajuato, 4 May 1594, Mexico City; *ibid.*, f. 2v, statement by Francisco Domínguez.
121. *Gobernantes del Perú, op.cit.*, vol. 11, pp. 55–7, Fernando de Torres y Portugal to the king, 25 April 1588, Lima.
122. AGI : L 129, 'Autos que se hizieron por el teniente de correg.$^{or}$ de Potosí cerca de otra nueva ynvención de metales que pretenden hazer en aquel asiento algunas personas', 1588, contains a copy of the viceroy's prohibition of 21 January 1588 and the arguments of Juan Pérez de Cuentas representing the interests of the mercury factors. Other documents relating to predicted consequences of Corzo's process are reproduced in Bargalló, *op.cit.*, p. 260f.
123. AGI : L 570, *lib.* 15, ff. 69v–70, the king to García de Mendoza, viceroy of Peru, 22 September 1590, Escorial.
124. Bargalló, *op.cit.*, ff. 401–3, reproducing the letter from Pedro Zores de Ulloa, *corregidor* of Potosí, to the viceroy.
125. AGI : L 273, ff. 294–7, Pedro Zores de Ulloa to García Hurtado de Mendoza, viceroy of Peru, 13 October 1590, Potosí.
126. AGI : L 570, *lib.* 15, ff. 107–8, the king to Hurtado de Mendoza, 9 October 1591, Escorial.
127. AGI : P 238i/10, Sancho de Valenzuela, *procurador general* of Potosí, petition on behalf of the town's residents to unnamed crown dignitary, (1596).
128. *Ibid.*, statement by Luis de Quinones Osorio, Potosí, 1596.
129. *Ibid.*, recommendation from Cristóbal Delgadillo, Potosí, 1596.
130. AGI : L 131, letters from Enrique Garcés to the king, 22 December 1590, and another undated, in an unfoliated group of documents relating to him. Garcés had returned to the Peninsula after 44 years in Peru, partly to present the king

with this project and also to seek reward for his past services. He said 1800 *quintales* of iron were being consumed in the amalgamation process each year at Potosí; whether this figure or the higher estimate given by Quinones Osorio six years later is considered, it amounts to heavy demand, from one quarter to one third of the average annual weight of mercury consumed there.

131. AGI :P191/17, ff. 9v–10, evidence of Gonzalo de Funes, mill-owner of Potosí, 16 June 1595, La Plata: *ibid.*, ff. 34v–6v, evidence given by Diego de Robles Cornejo, 22 June 1595, La Plata. The king's instructions called for information on the treatment of Indians at Potosí and a report on the processing of silver ore with iron; *ibid.*, f.1, the king to *licenciado* Bernardino de Albornoz, *oidor* of the *audiencia real* of Las Charcas, 22 June 1594, Escorial.
132. Lohmann, *op.cit.*, p. 155. The figures for annual mercury production at Huancavélica show a sharp dip in 1595, though the correlation is not simple: there were also years of high production (1591 was one) when Corzo's process was in use; Sanz, *op.cit.*, vol. 1, pp. 497–8 tabulates the data.
133. Both Funes and Robles Cornejo gave this explanation; *loc.cit.*, (note 131). Another explanation given by the *procurador general* of Potosí, *loc.cit.*, (note 127) supposed that the function of the iron was to remove sulphur, antimony and other impurities in silver ore which otherwise 'attenuated' the mercury.
134. M. Ulloa, 'Castilian seigniorage and coinage in the reign of Philip II', *Journal of European Economic History*, **4** (1975), 459–79.
135. *Ibid.*, and the same author's *La hacienda real de Castilla, op.cit.*, p. 427f.
136. C. del Rivero, 'El ingenio de la moneda de Segovia', *Revista de archivos, bibliotecas y museos,* **38** (1918), 20–31 and 191–206; **40** (1919), 146–150. The archduke arranged for the craftsmen to be supplied with interpreters in Spain who would also 'control' them because 'the German nation is given to drinking'; AGS : GA 145/48, Archduke Ferdinand to Baron Khevenhüller, Imperial ambassador in Madrid, 4 February 1582, Innsbruck. Neglect of the river bed was blamed for the flooding of the adjacent monastery of Santa Maria de los Huertos after the erection of the mint; ARP: *Administraciones Patrimoniales: San Ildefonso, legajo* 1, 24 October 1601.
137. *Cedulario indiano, op.cit.*, vol. 3, pp. 413–14, the king to the president and *oidores* of the *audiencia real* of Los Reyes, and the officials of the royal treasury in that city, 16 August 1563, Madrid.
138. Juan Fernández del Castillo, *Tratado de ensayadores* (Madrid, 1623), preface.
139. A. Matilla Tascón, *Historia de las minas de Almadén,* **vol. 1:** *desde la época romana hasta el año 1645* (Madrid, 1958) is the standard history.
140. *Ibid.*, p. 121.
141. Calculated from data in Matilla Tascón, *op.cit.* During Philip's reign Almadén sent 76 000 *quintales* of mercury to Seville for shipment to New Spain. In the same period the mines of Huancavélica produced around 166 000 *quintales* for consumption in Peru. When in 1593 the viceroy of Peru suggested sending Huancavélica's surplus to New Spain, replacing the shipments from Seville, the king replied that this was 'not desirable'; he had to consider the interests of Almadén and the contracts there with his creditors, the Fuggers; AGI : L 570, *lib.* 15, f. 154, the king to García Hurtado de Mendoza, 29 December 1593, Madrid.
142. AGS : *Expedientes de Hacienda, leg.* 537, the king to Alonso de Arellano, *alcalde* of Seville, 16 January 1560.
143. AGS : CJH 46/167, officials of the *Casa de la Contratación* to the king, 28 August 1562, Seville.
144. Matilla, *op.cit.*, pp. 70–3.
145. *Ibid.*, p. 100.
146. Ambrosio de Morales, *Las antigüedades de las ciudades de España* (Alcalá, 1575), p. 48.
147. Archivo Histórico Nacional, Madrid: *Ordenes militares (Archivo de Toledo), expediente* 37,888, unfoliated, January 1593.
148. IVDJ 22c/98b(v), Hans Schedler to the king, c. 1580.

149. Matilla, *op.cit.*, pp. 94–5.
150. *Ibid.*, pp. 95–7 and 116–17.
151. The document containing this evidence is now reproduced in full in G. Bleiberg, *El 'Informe secreto' de Mateo Aleman sobre el trabajo forzoso en las minas de Almadén* (London, 1985).

# 5

## The crown's interest in medicine

### *1 Royal hospitals for the poor, the sick and the workers*

Like the monarchs who had preceded him Philip II regarded charitable acts as an important part of his Christian duty towards his subjects. He therefore continued the giving of alms to the poor and the provision of assistance to the sick, one reason for the crown's frequent concern with medical affairs.

This policy was applied to all of his domains. In his African possession of Melilla, Philip, perturbed by reports of the decadent state of the hospital there for the poor, appointed a new administrator – there was no one in charge – to organise the supply of beds, clothes and alms, and made a grant of 50 ducats for medicines, stipulating that these were to be stored in a place where they would not be spoilt.[1] In the New World, soon after the first conquests, Isabella had ordered the foundation of hospitals for the sick poor, both Indian and Spanish; and in 1541 Charles V extended this policy to his greatly enlarged American empire, instructing viceroys to be diligent in establishing hospitals in all *pueblos* so that Christian charity and medical treatment could be given to the sick poor. By Philip II's reign there was a crown plan for the design of new settlements which specified particular locations for the sites of official municipal buildings, customs houses and hospitals: the hospital for the poor and the sick was to be put next to the church; but for those with contagious diseases a separate hospital was to be built on high ground, isolated from the main centres of population and where no wind might blow infected air back to the town.[2] Notable royal foundations in New Spain included the Hospital Real de Naturales for poor and sick Indians of Mexico City. In response to local requests Prince Philip approved the construction of the hospital in 1553, authorising the use of royal funds for the building costs (4000 ducats were released for this in 1553–6) and an annual endowment of 400 *pesos*;[3] it continued to function until 1822. Philip also reacted positively to the request for a hospital for sick mulattos in Mexico City; after the matter was discussed in the Council of the Indies, he instructed the viceroy to consider how grants of land and rent could be used to endow the hospital 'for the service of God and ourself'.[4] By far the greatest number of hospital foundations in sixteenth-century New Spain were due to private individuals; but sometimes these passed to royal control. This occurred with two of the numerous hospitals created by Bernardino Álvarez, founder of the pious Order of Charity. His Hospital de Nuestra Señora de la Consolación in Acapulco was established (c.1580) to care for sick seamen, caulkers and slaves arriving at the port; in 1596 Philip provided funds for its maintenance and it became a royal hospital. The same happened with Álvarez's Hospital de San Martín on the isle of San Juan de Ulúa, one of the principal harbours for the Indies traffic; Philip maintained it, and the brothers of the Order of Charity cared for the sick.[5]

Royal hospitals were part of the royal patrimony, and during Philip II's reign a more vigorous defence of crown rights over them is discernible; this was another manifestation of that reassertion of regalia shown in Philip's incoporation of mines. In Mexico City care was taken to exclude the archbishop of Mexico from the government of the Hospital Real de Naturales; he could not even visit it unless accompanied by the viceroy or other royal official. The same jealousy over jurisdiction was exhibited in the control of Spain's leper houses, which since the Catholic Monarchs had been regulated by the crown. Under Charles V they had been visited by royal officials, the *corregidores*, together with prelates; but Philip subsequently refused this ecclesiastical intrusion and in the dispute which ensued at Toledo in 1591 the king's lawyer successfully defeated the archbishop of Toledo's groundless claims.[6] A spirit of greater cooperation between church and crown was responsible for the more efficient organisation of the Peninsula's other hospitals brought about by the movement for consolidation in the later sixteenth century. In this way the numerous small medical foundations, sometimes so inadequately endowed that they could barely pay their staff, were merged into general hospitals in Madrid (1587), Valladolid, and most dramatically of all in Seville

(1584–9) where the resources of 75 hospitals were pooled to form just two large centres for the sick.[7]

The crown also provided infirmaries for workers employed in its service. At the royal arsenal in Barcelona there is no record of special provision for sick or wounded craftsmen, probably because the city's facilities were used. But in smaller or remote localities workers usually received the benefits of royal welfare, whether out of charitable sentiment or in the interests of efficient production. At Guadalcanal the royal administrators of the silver mines found that the four hospitals in the surrounding area were inadequately equipped to deal with the diseases or accidents of the miners. The princess regent therefore approved the employment of Francisco Rodríguez, a local physician and surgeon, with an annual salary of 24 000 *maravedís* to treat patients and, with his wife's help, to care for them in his house. This was preferred to the foundation of a special hospital for reasons of expense. Soon the crown authorised payments to extend the physician's house to accommodate more patients. By the end of 1558 this makeshift infirmary had 12 beds and a pharmacy stocked with medicines bought in Seville. The physician himself functioned as the apothecary; he was also said to be good at blood-letting. Throughout the 20 years of royal administration a physician attended to the needs of Guadalcanal's workforce. From 1562–75 Pedro de Paz visited the site daily and answered night calls, coming by mule from his home two leagues away, grumbling that he had to pay out of his own pocket for the use of the animal and for the boy who looked after it.[8]

The remote and unhealthy royal mercury mines also required special medical services. At Huancavélica a hospital was created for the treatment of Indian miners, and in the contract made with the mine operators in 1577 the crown agreed to maintain the hospital with payments of 2000 *pesos de plata ensayada* (2400 ducats) each year.[9] And at Almadén the king's contracts with the Fugger lessees required them to use income from fines imposed for thefts of mercury to establish an infirmary for workers 'who usually become ill there', providing beds, medicines, a physician and a barber. From a lawsuit in December 1589 in which Lupercio de Viveros was acquitted of charging fees for treating convict workers at the mine, we learn that this physician had been engaged by the Fuggers the previous year with a salary of 39 375 *maravedís* to attend to all on the site: the royal officials and their families, the foremen, craftsmen, convicts and slaves.[10]

The king's attitude to the health of Indian workers is most clearly revealed in his decisions on the cultivation of coca in the Andes. The plantations were situated in a humid zone, thought to be unhealthy and the cause of 'an incurable disease worse than the pox'.[11] Viceroy Toledo sent Alegria, a physician, to the area to investigate the health hazards.[12] The plantations were not the crown's but belonged to Spanish settlers in Cuzco; they were becoming wealthy by preying on the Indian addiction to the cocaine contained in the leaves of the plant, the habit of centuries and one that was sanctioned by ancient religious rituals. But the crown had an interest because the use of the drug by Indian miners at Potosí was supposed to assist silver production; the physiological action of the cocaine certainly deadened the sensations of hunger and thirst and, it was alleged, made it easier for the Indians to tolerate the working conditions. How did the king react to this? The silver mattered most to him; and while conceding that 'a host of Indians perish because of its cultivation in hot unhealthy regions' and 'although we have been asked to prohibit it, we do not wish to remove this assistance to work even though that may be imaginary' and, in the opinion of some, 'an illusion of the devil'.[13] Philip therefore allowed cultivation to continue, seeking to mitigate the health risks which that incurred by ordering all workers to be supplied with a change of clothing – working in wet clothes, due to the local humidity, was said to be one of the causes of the high mortality. The same royal decree compelled the plantation owners to employ physicians, surgeons and apothecaries.[14] Viceroy Toledo was ordered to implement these and other measures, such as the restriction of working hours and the prohibition of forced labour. He accordingly ordered the coca lords of Cuzco to donate one per cent of the proceeds of their crop to support 'the hospital of the Andes', and to send there any Indian worker who became sick or face a fine of 100 *pesos*. And in case the coca estate was far from the hospital or the illness too acute for the worker to be moved, all proprietors were instructed to supply their plantations with 'lancets for bleeding, oil, and corrosive sublimate for the treatment of sores'; if they were found to be without these a fine of 50 *pesos* would be imposed[15]. As for the use of the drug Philip's only objection was that it was associated with the idolatry of the Indians' primitive religion. He therefore urged prelates to be vigilant and prevent any applications in furtherance of superstitious rituals. In the interests of the true religion he forbade the plantation owners employing Indians on Sundays and holy days in drying coca leaves, because that deprived them of mass and the benefits of preaching.

A royal hospital for workers was also specially created during the building of the Escorial. Its constitution required the appointment of a resident apothecary to prepare purges and distil essences; and, particularly interesting for revealing contemporary attitudes to hospital care, much attention to preaching and praying 'because it matters more to cure the soul than the body'[16].

But the number of royal hospitals in Spain and the Indies was very small compared to those founded by individuals and religious. Whatever the centralising aspirations of Philip II's government, there was no Council of Health and the bulk of Spain's seven millions received no direct medical assistance from the crown. In the countryside the sick were attended by the clergy or itinerant healers. In the towns public health had been entrusted to the municipal authorities since the Middle Ages and remained a guarded preserve of their autonomy along with the control of markets, prices, and provisions. The towns administered hospitals, contracted physicians[17] and arranged for the free treatment of the poor. And in Valencia, where public health was most developed, special officials were responsible for hygiene and the prevention of the adulteration of food[18]. The terrible epidemics which struck Spain throughout the reign were left largely to the local authorities to deal with. The crown's role was advisory and financial. In 1582 the city of Seville, still in the grip of the pestilence which had broken out the year before, appealed to the king for funds. The city had spent 86 864 ducats in fighting the disease after receiving assurances that it would be reimbursed by the crown; but only 24 000 ducats had been granted leaving 50 000 owing to physicians and other personnel employed. The crown's response was to order an inspection of the accounts, to which the city replied that this would cause delays, assuring the king that no fraud had been committed and that all of this large sum was unavoidable expenditure authorised by the count del Villar, *asistente* (the king's representative) of Seville[19]. Reports by the *asistente* on the course of the epidemic went to the Council of the Indies, and after hearing that in a few months 1276 had died from the contagion, it advised the king to consult one of his most valued physicians, Francisco Valles[20]. This was soon done; Valles and another king's physician, Antonio Fernández de Vitoria advised 'all means possible to be used against the pestilence because of the great number of sick and dead'. Their advice, no different from what would have been given elsewhere in Europe, was based on the contemporary belief that plague was caused by the corruption of the air. They recommended the lighting of fires in all streets, burning aromatic wood, 'even though this is very expensive', to correct the air, and the use of large quantities of vinegar for the same purpose. The city should ensure that pharmacies were well stocked, if necessary making the public pay for medicines dispensed. The medicines recommended were those commonly used against plague: bezoars (stony concretions formed in the stomachs of goats and other animals) supposed to possess virtues as antidotes; and those elaborate concoctions of multiple ingredients, mithridatum and theriac, 'the best medicine there is'. The usual precautions were indicated: the importance of burning rubbish and infected clothing; the isolation of the sick; the need for a healthy diet – Seville's officials were advised to ensure the soundness of fish and pork sold to the public, and to reduce sales of fruit susceptible to rapid decay. The gathering of crowds was to be avoided, and though 'above all else there should be much praying' (an allusion to God's wrath as the ultimate source of the visitation), 'this should be done without bringing many people together'. As a further precaution against spreading the contagion they advised the king not to receive any letters from the afflicted city.[21]

The crown acted in similar ways during the much more serious pestilence of 1597–1602 which affected the entire Peninsula and caused over half a million deaths. In his study of this Bennassar found that the crown gave the *corregidores* of Burgos, Toledo, Seville and Valladolid no more than moral support, encouraging or congratulating their efforts; but that Santander, a coastal town of strategic importance, received more substantial assistance: a paid physician and exemption from the *alcabala*, the sales tax normally levied for the crown, to stimulate resettlement in a town where five sixths of the population had died.[22] During this outbreak Philip II ordered Luis Mercado, his physician, to write a treatise on the plague. This was published, first in Latin, in 1598, and then a year later, at Philip III's request, in Castilian 'so that in all provinces, cities, towns and localities the disease is recognised, and it is known how healthy places are to be protected and how in places afflicted it is to be overcome, and what each must do to protect their health and what remedies are to be used to treat the victims.'[23]

A more determined intervention by Philip II in the health of towns was occasioned by moral as well as medical considerations. In 1570 he issued regulations for brothels in Seville, requiring a physician and surgeon to visit them once a week, sending all diseased prostitutes to hospital for immediate treatment. His order for brothels to close during Holy Week was ignored. The vice of prostitution and the accompanying risk of the spread of syphilis in Granada, Madrid and other cities were of concern to the king on several occasions in the 1560s and 70s.[24] The matter does not seem to have caused any important conflict between the crown and the municipal authorities; conflict did arise however from the crown's continuing attempts to control medical practice.

## 2 *The king's control of medical practice*

There was no dispute with the towns over the training of physicians and surgeons. In fact the crown's intervention in university medical education was the direct result of petitions made to the king by the delegates of the towns of Castile during the meetings of the *Cortes* which assembled to approve the king's taxes. In 1555 those delegates, the *procuradores* as they were called, voiced the towns' grievance that the lives of people were put at risk by the treatment given by imperfectly trained practitioners; the alleged cause was fraud in obtaining degrees as students moved from one university to another bringing false statements of their graduation in arts. The *procuradores* asked for the prevention of this abuse and for the introduction of a new regulation requiring medical graduates to serve two years' apprenticeship with a qualified physician before being allowed to practise. In response to this Princess Juana, Charles V's regent, ordered that henceforth authenticated documents signed by university rectors and notaries would be the only acceptable evidence that a student had really studied the courses as he alleged.[25] Philip II went further than this to meet the towns' demands. In 1563 at the *Cortes* convoked in Madrid he decreed that the requirements for a bachelor of medicine were graduation in arts at an approved university (in Castile that meant Salamanca, Valladolid or Alcalá) and a further four years' study of medicine; to be followed by two years' training with an experienced physician before practice was permitted[26].

Enforcement of this proved troublesome and after continuing complaints in the *Cortes* that the universities were neglecting to implement the two years' post-graduate apprenticeship, all Philip could do was to insist that testimony of its satisfactory completion was presented to the municipal officials of the town where the doctor intended to practise, and introduce a penalty of eight years' suspension from practice for infringements.[27]

The cities also pressed the crown to improve the training of surgeons. In the assembly of the *Cortes* in 1551 the *procuradores* had complained of 'a great shortage of surgeons' in Castile, attributing this to the neglect of anatomical dissection in the universities. On that occasion the royal reply was that 'we have already arranged for universities to teach and perform the necessary anatomies'.[28] While there was some truth in this – in 1539 Charles V had authorised the donation of corpses of criminals and from hospitals to the university of Alcalá, and some anatomy was being taught at Valladolid in 1550 using corpses – the *procuradores* were justified in their assertion that Castilian universities were backward in this respect, especially when compared to Aragón; Valencia had one of the earliest chairs of surgery in Europe (1502) and there the new Vesalian anatomy had soon been taught by Pedro Jimeno (from 1547) and Luis Collado (from 1550) who had held the chair in surgery[29]. Again Philip II took more positive action. In January 1566 the municipal council of Salamanca had written to him that the continuing shortage of competent surgeons had resulted in many deaths because of the improper treatment of minor wounds; it appealed for the creation of a chair of surgery in the university. After discussion in the royal Council of Castile Philip wrote to the rector of Salamanca asking the university to consider the proposal. The professors of medicine opposed it alleging that there was no neglect of anatomy in the faculty; but the university as a whole voted in favour and the following year Andrés Alcázar was appointed to the new chair. He held this position until his death in 1584, and made important contributions to the diagnosis of head injuries and to the technique of trepanning.[30]

During the 1580s the *Cortes* again called for chairs of surgery to teach the art properly and avoid the current application of 'violent and inappropriate' remedies by ignorant practitioners.[31] But there was no further initiative from the crown until 1593–4 when Philip wrote to Alcalá and Valladolid (and to Salamanca, indicating that the established chair there had not continued to be filled) that 'it would be to the benefit of these our kingdoms if there were a course in surgery to train those who are to practise this art'. The king's letters, conspicuously seeking to raise the standing of surgery and to counter academic disdain, announced that the occupants of chairs of surgery would be accorded the title 'surgeon royal', and students satisfactorily completing surgical studies would be graced with the title *licenciado* like the successful graduates in law or medicine.[32] By the end of 1594 two new chairs were filled: Francisco Ruiz at Valladolid and Luis de Victoria at Alcalá. Only in Italy had surgery made such inroads in the university world. Previously Philip had required surgeons to have four years' practical training in a city hospital; now they had to study some arts courses at university (but without graduating), medical and surgical courses, followed by two years' practical experience. But the insistence on univesity training for surgeons before adequate numbers of students had come forward created difficulties and ten years later, after reports in the *Cortes* that 'many surgeons' had died in the calamitous plague, Philip III revoked the regulation, and surgeons without university training were readmitted.[33]

Unqualified empirics and healers served a useful function at a time when there were not nearly enough physicians or surgeons to cover the Peninsula. These men treated cataracts, ring-worm, dislocations, hernia and the stone, just as their counterparts did in contemporary England. They were welcome both to the crown and the towns. In 1588 Philip confirmed their rights to treat these disorders, although for hernia and the stone the presence of a physician and surgeon was required.[34] One of them, Francisco de Somovilla, employed to treat the royal family for the stone, hernia and dislocations, served from 1567–74 with a salary of 40 000 *maravedís*, rising to 60 000, no less than most of the qualified royal physicians received. He was succeeded in the same capacity by Diego de Somovilla, perhaps his son. And in the *Cortes* of Castile in 1579 the *procuradores* voted to engage the empiric Agustín de Alba to spend one month in each of their towns and cities, teaching surgeons his techniques of curing urinary disorders and excrescences of the flesh.[35]

But the greatest stir was caused by the Vizcayan empiric Aparicio de Zubia who claimed to have discovered a marvellous medicinal liquor, a herbal composition effective in various diseases and especially for wounds, avoiding the torments of surgical instruments, blood-letting and purges.[36] Around 1550 he had used his oil in the city of Granada with such effect that he had become known as the 'saint of Vizcaya'. But his success aroused the opposition of physicians and surgeons and he was twice arrested in the city for unlicensed practice. In 1552 the crown intervened with Charles V's authorisation of the use of the oil, but the resentment of the physicians continued to obstruct. Zubia crossed to Flanders to petition Philip, at the same time treating wounded soldiers on the battlefield of St. Quentin. In January 1559 Philip ordered an examination of the oil 'to end the tribulations and calumnies' of the physicians. The bishop of Guadix, administrator of the *Hospital Real de Corte* (an itinerant hospital which attended courtiers), observed the beneficial results of the oil: 'those wounded in the head and arms recovered within five or six days without the usual restrictions on diet, without fever, without blood-letting or amputation, and without the exhorbitant fees of surgeons'. He and the deputies of the hospital unreservedly recommended the oil and advised the king to support Zubia and encourage him to reveal the secret of its composition. The *procuradores* were equally impressed. Gathering in Toledo for the assembly of the *Cortes* in 1560, they accepted Zubia's invitation to witness the treatment he was giving outside the city hospital. They found the oil 'very advantageous' and proposed that Zubia be engaged to impart the recipe, indicating which herbs to use and the method of preparation to a dozen persons in Valladolid, from which centre with its 'great concourse of people from various parts' others could communicate the information throughout Castile. Zubia's widow subsequently rejected an offer from the crown of 30 000 *maravedís* a year to reveal the secret as inadequate for 'a universal remedy'. But in the end an agreement must have been reached because the 'oil of Aparicio' became extensively used. The documentary evidence shows that 20 *libras* of the oil were sent in 1584 to the Azores for the infantry garrisoning Terceira after the recent conquest of the isle.[37] And in 1591 out of 52 vessels of an armada no fewer than 51 were supplied with this oil,[38] clearly indicating that Zubia's remedy had come to be accepted as a standard medicine and that the opposition of jealous physicians had been overcome.

There was no quarrel between the crown and the cities over the occasional employment of unqualified practitioners like Zubia. Nor were they at odds over the exclusion of other medical practitioners on the grounds of race. At the beginning of the century the Catholic Monarchs' test of purity of race had been introduced to keep those of Jewish and Moorish descent out of numerous occupations including medicine and surgery. The fears of what these despised groups might do to the Old Christian community if they were ever allowed to infiltrate the ranks of physicians and apothecaries continued to be expressed in exaggerated outbursts. Members of the clergy were in the van of the persecutors. In the 1580s Andrés de Noronha, bishop of Plasencia, was urging Philip to enact stronger measures to prevent *conversos* from occupying positions of trust like that of physicians.[39] He believed he had evidence of a conspiracy between the Jews of Portugal and those of Constantinople[40] for the destruction of Christians through medical treatment. Noronha, himself Portuguese, had been given letters, a supposed correspondence between the plotting Jewish communities, by the tribunal of the Spanish Inquisition at Llerena. One of the letters, purportedly written by Portuguese Jews, complained of persecution and announced their plan to achieve liberation by 'a single remedy which was to teach their sons the science of medicine and the art of the apothecary so that these might become the means of exterminating their persecutors'. The reply, allegedly from a Jew of Constantinople, encouraged them 'to persevere with their stratagem of teaching these sciences to their sons' and in addition advised them to become lawyers 'in order to take estates from those who are not of their race'. Noronha had been given the letters at the time when he held the Portuguese see of Porto Alegre and before Philip's annexation. He had at once communicated the information to King Sebastian who 'took my petition so much to heart' that he took appropriate action to protect medical studies at Coimbra.[41] And now 15 years later he wanted Philip to show the same vigilance.

Philip had already given his support to measures intended to achieve precisely the same ends. In 1564 he reaffirmed royal approval of the constitution of the fifteenth-century College of Apothecaries of Valencia whose syndics continued to insist that 'it is desirable for the well-being of the city of Valencia that apothecaries who prepare medicines are not of Jewish descent' and accordingly to 'request Your Majesty in the interest of greater security that *conversos* may not be apothecaries, make medicines, hold a pharmacy nor make nor sell things related to pharmacy, nor be admitted to the examination, under penalty of 500 ducats and permanent expulsion from the city'.[42] The regulation of 1529 which disqualified anyone who married a woman of Jewish descent from becoming an officer of the College was also ratified by Philip. And in Barcelona, where for at least 50 years all who intended to practise pharmacy had been required to present proof of *limpieza de sangre*, the viceroy of Catalonia in 1562 ratified this statute which was part of the constitution of the city's College of Apothecaries.[43] The story was the same in other cities. Zaragoza's College of Apothecaries excluded *conversos* and Moriscos for the greater safety of the city, and in 1591 the municipality of Seville[44] maintained the same restrictions.

The crown, the municipal authorities and the guilds which they controlled, everywhere favoured the policy of preserving the medical arts for Old Christians. But there were loopholes which it was not in the interests of the crown to close at a time when the Peninsula needed all the skilled practitioners who could be found, especially the *converso* physicians who were among the best. The dilemma was appreciated by Juan Guardiola, a Benedictine. He congratulated the university of Barcelona for its assiduous inquiries into the ancestry of those aspiring to the degree of doctor in medicine and for rejecting as unworthy of such office any found to have the blemish of Jewish descent. And contrasting medieval and present practice in the appointment of royal physicians he reflected that while Alfonso the Wise had only required his physicians to be wise and experienced and not necessarily of good lineage, Philip II generally refused physicians with racial blemish, although occasionally some were admitted because of their talent 'in which case it is licit to take advantage of this'.[45] Of the distinguished physicians and surgeons who rose to the service of Philip's household Cristóbal Pérez de Herrera and Dionisio Daza Chacón are said to have been of Jewish descent.[46] Even the Inquisition in times of need agreed to employ a *converso* physician. Its hypocrisy is evident in its reply given to the request from the tribunal of Logroño in the 1570s. That tribunal wanted to consult a skilled physician but could not find one of pure race. Might it be permitted to use the services of Doctor Bélez, a *converso*? From Madrid came the decision of the *Suprema*: he could be consulted but without recognition of his title.[47]

A few Morisco practitioners also managed to surmount the obstacles. On two occasions Philip brought in Morisco healers, Jerónimo Pachet from Gandía and Pinterete from Valencia, to treat his ailing sons after the royal physicians had failed to cure them. Some Moriscos graduated in medicine at Valencia and Granada. In 1550 one of the 35 physicians practising in the city of Valencia was a Morisco;[48] and at Llerena the census of 1594 shows that an apothecary and a physician were Moriscos.[49] But, these exceptions apart, Morisco practitioners remained a persecuted minority; shortly before their expulsion from the Peninsula (1609–11) rising fears were expressed in the *Cortes* that Moriscos were infiltrating the medical faculties of Castile, putting them in a position 'to kill more of this kingdom than the Turks and English'.[50]

While there was no difference between the crown and the towns on the training of medical practitioners, a serious conflict arose over their licensing. Licensing of physicians, surgeons and apothecaries was one of the privileges which had been granted to the towns of Castile and Aragón by monarchs in the thirteenth and fourteenth centuries. With royal approval guilds of apothecaries had been created in numerous towns under the supervision of the municipalities; they controlled all matters relating to the practice of pharmacy in the locality. And local justices issued licences to town physicians and surgeons. The tenacity with which the towns would defend these privileges had already been manifest in the fifteenth century when first Juan II of Castile, and then the Catholic Monarchs, tried to recover the powers of licensing for the crown by appointing physicians for this purpose. Juan had been told by the cities that his *alcaldes examinadores* contravened the rights and customs of the towns; he had retracted but soon tried again. The similar *alcaldes examinadores mayores* of the Catholic Monarchs, introduced in 1477 to inspect and license 'physicians, surgeons, healers, apothecaries, spicers and herbalists', had aroused municipal opposition in Segovia, Jerez, Córdoba and Seville. Apart from resentment of the crown's intrusion there had been allegations of profiteering by these new officials, selling licences to the inept. There were also complaints that the officials had sent surrogates who had accepted bribes for the issue of licences. The use of surrogates was inevitable since with just a few officials – only four were appointed at first – Ferdinand and Isabella had hoped to control licensing throughout Castile; their response was to prohibit the use of surrogates. Something similar to this was experienced in England where the Royal College of Physicians, created in 1518 to regulate the practice of medicine, had attempted to cover the country from London by delegating its authority to representatives in the provinces; that too had resulted in the abuse of selling licences for profit. That College had been forced to seek the assistance of Justices of the Peace and mayors for licensing in localities far from London; the difference was that England did not have the numerous guilds of physicians and apothecaries which were established in Castile and Aragón, some powerful enough to obstruct any plan for licensing by a central authority. In recognition of the practical difficulties and out of deference to local complaints of 'harassment', Charles V issued a decree (1523) which confined the activities of the royal officials – they were now called '*protomédicos*' – to a radius of five leagues from the court, and for the rest of Castile left the licensing of practitioners and the inspection of apothecaries in the hands of municipal officials and local physicians. This was re-enacted by Philip in the *Cortes* in 1567[51].

But the stability of this arrangement was threatened in two ways. Although the *protomédicos* were forbidden to summon anyone from outside the five-league limit, would-be practitioners came of their own accord to Madrid to seek licences from them. In July 1579 Pedro de Camaño travelled from Galicia, at the periphery of Castile, for this purpose; he wanted to practise surgery in that region. Whether this had been refused by local officials there is not recorded. In Madrid he was interviewed by Philip's *protomédico*, Diego de Olivares, and three other qualified practitioners. They asked him questions on 'anatomy of the human body, wounds, ulcers, sores and other abscesses' and, satisfied with his replies, they gave him a licence to practise surgery 'in all the kingdoms and domains of His Majesty'.[52] It contained a clause warning local justices, with threats of heavy fines, not to prevent him practising his art. On this occasion the municipality of La Coruña recognised the *protomédico's* licence and Camaño was soon practising in that city. But there was clearly a potential source of conflict here and future research may well reveal cases where individuals arriving in cities with a *protomédico's* licence were not so favourably received.

More serious and well-documented opposition arose from the crown's attempts to impose *protomédicos* in the kingdoms of Navarre and Aragón, where these officials were the unwelcome agents of a foreign crown, apart from the challenge to established privileges of the guilds. Navarre had become a part of Castile in 1515 soon after Ferdinand's annexation of the territory as a security against French invasion. During the following decades its government had been reorganised with the creation of the royal Council of Navarre and the appointment of viceroys. But local tradition remained strong, and the kingdom retained its own law, parliament and the privileges of its municipal institutions, among which were the guilds of physicians, surgeons and apothecaries of Pamplona, Tudela and Estella. Some time after 1527 Charles V decided to appoint a *protomédico* of Navarre 'with all the honours, privileges and powers enjoyed by our other *protomédicos* of our kingdoms of Castile'. The crown was sufficiently sensitive not to appoint a Castilian; the first occupant, Dr de Santa Cara, was a Navarrese physician from Olite. He subsequently had found it necessary to request the viceroy of Navarre to order justices and other officials throughout the kingdom to recognise his authority. But this had no lasting effect; his son who succeeded him also complained to the crown that his orders were being ignored; all the crown could do was issue orders (October 1548) for cooperation and threaten fines for disobedience.[53] Philip continued to appoint men to the office; in January 1572 he approved a salary of 15 000 *maravedís* for Dr Calduendo who had indicated the 'considerable labour' of personally inspecting all the pharmacies of the kingdom and penalising 'offences and errors' of medical practitioners.[54] In May 1581 Dr Tarazona held the position. After the viceroy and the Council of Navarre had received his report on medical matters requiring correction, a letter in the king's name was circulated throughout Navarre. This ordered physicians to comply with the regulations requiring them to advise the dangerously ill to make preparations for confession; to visit the poor whenever summoned and without charge; to avoid disputes on the cost of prescriptions by observing the agreed tariff (in May 1574 the crown had prepared an official tariff fixing the price of medicines in Pamplona; prescriptions had to be dated and had to indicate the cost of medicines prescribed); to provide signed instructions in cases of blood-letting, authorising surgeons to take a specific quantity of blood; apothecaries were to refrain from dispensing pills, purges and medicines without the authorisation of a physician, and to prepare all internal medicines 'with their own hands and not to give this to a servant to do'.[55] But local resentment over the crown's attempts to control practice were not readily overcome. In 1586 the physicians of Pamplona were objecting to the activities of Lazcano, the new *protomédico*. There was a dispute over the licensing of a physician who was applying to practise in Estella. The *protomédico* and *licenciado* Bayona, a local physician, interviewed the aspirant in Pamplona and disagreed on his suitability; Bayona advised rejection but was over-ruled and he subsequently complained that in this and other cases the *protomédico* was ignoring other opinions and acting as if he were alone. It had been a manoeuvre frequently adopted by the towns of Castile to allege that *protomédicos* were licensing the inept. Bayona and four other local physicians also protested over the *protomédico's* intrusion in another former local preserve: the inspection of apothecaries and their shops. Fees were charged for these tours, apothecaries paid for the inspection and official approval of their stocks, which sometimes involved the removal and destruction of stale medicines; the fees covered the costs of the *protomédico* and the notary who accompanied him.[56] But according to the physicians of Pamplona, who in the past had exercised this function with some remuneration, the *protomédico* was over-charging. His visits were moreover said to be ineffective because as soon as news of his departure from Pamplona reached the apothecaries of Navarre they began to prepare for the inspection 'loaning medicines to one another to fill their pharmacies, and as soon as the *protomédico* leaves they return to their former neglect'. Therefore it would be better for each local council to employ its local physician, and for an apothecary from another district to perform the inspection without alerting apothecaries of the impending visit to their premises. This the physicians said was the custom of Navarre and it should be observed.

This petition went to the *Cortes* of Navarre and was largely upheld. The *Cortes* requested the viceroy to order the *protomédico* to issue licences only with the participation of local qualified practitioners; a concession to the *protomédico* permitted him to nominate his collaborators. As for the inspection of pharmacies, the *Cortes* was adamant that former custom be upheld and the *protomédico* excluded.[57]

In neighbouring Aragón the *protomédico* is a shadowy figure and it is the guilds which stand out as the dominant powers. In Zaragoza the College of San Lucas, San Cosme and San Damián, the guild of physicians and surgeons, so effectively controlled practice in and around the city that there was no place for the king's man to function. Its constitution, granted by previous monarchs and elaborated in the sixteenth century, established the right of officials of the College to restrict medical practice to physicians who had been examined by them and issued with licences bearing the seal of the College; empirics like tooth-drawers and healers of the stone could also practise after receiving a College licence. The College even asserted its powers beyond the city, issuing licences to surgeons in other towns and localities of Aragón. The surgeons whom it examined were expected to know Latin well enough to be able to reply in that language to questions put to them. The College also appointed men to inspect apothecaries' premises, prescriptions and medicines to ensure that medicines were being correctly dispensed by qualified practitioners.[58] The only concession made to the *protomédico* by the *fueros* (laws and privileges) of Aragón was that he might examine and license outside of the city of Zaragoza.[59] But that meant the countryside because licensing in the other cities of the kingdom was similarly controlled by fifteenth-century guilds in Huesca, Tarazona and Calatayud.

As for Catalonia much more is known about the functioning of the king's superintendent of medicine in a land of jealously guarded privileges from the thesis of Jordi González.[60] It shows that much of the friction was removed by a compromise in which the privileges of the guilds were generally respected and some collaboration achieved between them and the *protomédico*. The crown could hardly ignore the *fueros* which guarded the liberties of the Principality and guaranteed the privileges of the municipal guilds. Only in the eighteenth century, when these *fueros* were annulled as a punishment for Catalan disloyalty during the Bourbon succession, would the ground be cleared for the establishment of an all-powerful crown superintendency of medical practice. But in the sixteenth century the crown's efforts in this direction were bound to be frustrated. Therefore Charles V's regret that the authority of his *protomédico* in Catalonia was not receiving due recognition.[61] Philip's acts reflect acceptance of the existing powers of the guilds in their moderate assertion of royal controls. The appointment in 1588 of Jerónimo Mediona (a Catalan physician like others who held the office) as *protomédico* of Catalonia gave him the autority to examine and inspect only those apothecaries in the Principality who were not members of guilds and therefore already licensed by them.[62] At the *Cortes* which Philip attended in Monzón in 1585 the king had attempted to check the guilds' control of pharmacy by ordering all inspections of apothecaries' stock to be done jointly by the *protomédico* with locally appointed physicians and apothecaries, removing unsound medicines 'irrespective of any privileges granted by the guilds'. But at the same *Cortes* he conceded full rights of inspection in Gerona to the town's college of apothecaries, and excluded the *protomédico* from sending anyone to perform the inspection.[63]

But it was the city of Barcelona with its fifteenth-century college of apothecaries which most successfully resisted the intrusion of royal power. This had been the institution which had drawn up the Peninsula's first official pharmacopoeia (1511).[64] The *Colegio de Boticarios* had complete control over pharmaceutical practice in the city; its officials, the *cónsules*, examined apothecaries after eight years' apprenticeship for admission to the College; fixed the prices of medicines; inspected the city's pharmacies and imposed fines for breaches of the College regulations. All of this Philip ratified; the capital was no place for the *protomédico* of Catalonia. But in the countryside, Jordi has shown that the control of pharmacy was the joint work of the *protomédico's* deputies – Catalan physicians – and members of Barcelona's College; together these representatives of royal and urban authority performed itineraries of inspection; there is record of visits in the 1590s to over 150 apothecaries in the small towns of the principality.[65] In the visit to Besalú in 1594 numerous rotten medicines were thrown onto the floor of the apothecary's shop for subsequent destruction, and the practitioner summoned to appear before the *protomédico* in Barcelona within 15 days. It was in the countryside that the *protomédico* of Catalonia was able to exercise most authority; free from the hold of the guilds, numerous apothecaries in rural localities were examined and licensed by the king's official. The prolonged dispute which began at Vich in 1602 between Mediona and eight of the town's apothecaries may or may not indicate a clash between the *protomédico* and the local guild – the cause is not clear[66] – but the underlying tensions of contested authority are still evident in the control of medical practice in seventeenth-century Catalonia.

No study has yet been made of the *protomédico* in the kingdom of Valencia. An authority on the period has suggested that it was only an honorary title there.[67] But was this the case? When in 1576 Philip appointed Luis Collado, Vesalian anatomy professor at Valencia, '*protomédico* and royal supervisor of everything concerning the art of medicine, surgery and pharmacy in the city and kingdom of Valencia', it was in terms which have the ring of more than the bestowal of a grandiose title. The *fueros* and privileges of the kingdom, city and university were to be respected; but Collado was meant to be active. He was instructed to inspect the city's pharmacies together with two apothecaries selected by the municipality, establish a tariff for medicines, and impose fines for offences, taking from this source his payment of 20 *reales* for each day's inspection. Officials in the kingdom were ordered to recognise his authority. And when Collado died in 1589 Philip lost no time in replacing him with the Paracelsian Llorenç Coçar.[68]

Philip's instructions may have been unrealisable because Valencia's fifteenth-century *Colegio de Apotecarios* had long exercised these powers in the city and beyond. When he attended the *Cortes* at Monzón (1585) the king granted the College jurisdiction over apothecaries throughout the kingdom, in response to its request that in visits of inspection 'neither the *protomédico* nor any other may intervene in any manner, because apart from this never being the custom, it is unnecessary'.[69] Exactly how this worked out in practice remains to be seen. Local historical studies are needed to investigate this and the licensing of physicians and surgeons outside of the city – within the city inspection and licensing were firmly controlled by the municipality. It is part of the much wider subject of the maintenance of urban autonomy in the centralising monarchy of Philip II, on which more research is needed.

In the last decade of the reign the *protomédicos* of Castile were given greater powers and their influence was felt beyond the narrow five-league belt around Madrid. In 1588 practitioners seeking

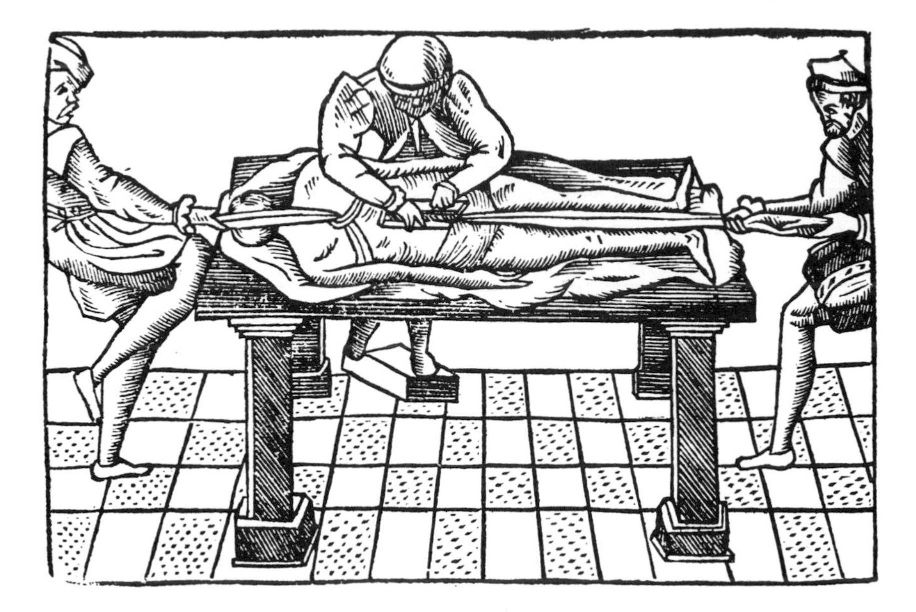

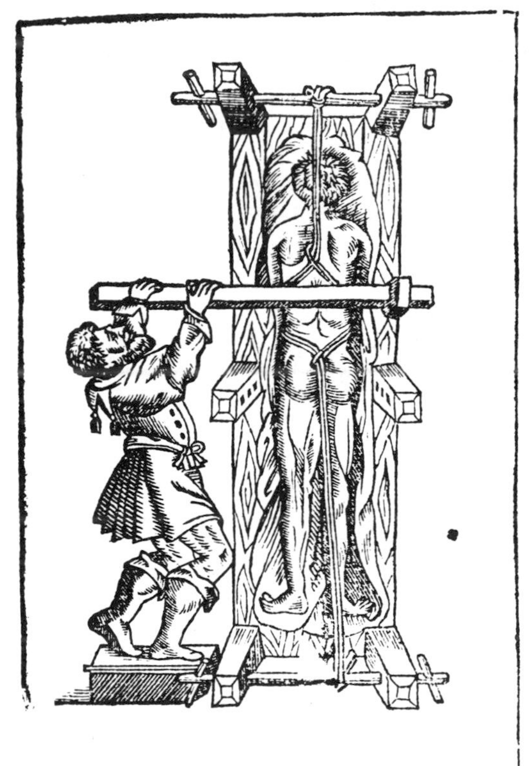

licences in the area of Madrid were subjected to a theoretical and practical examination by the *protomédico* and three examiners.[70] Physicians were questioned on a medical text and then went with the examiners to the general hospital of Madrid where they took the pulse of patients, were asked to give diagnosis and prognosis for each, and to indicate the treatment which they would have recommended. The examiners then reassembled at the *protomédico's* office and discussed the applicant's performance. Those wishing to practise surgery were also tested on theory and practice, the examiners observing their skills in applying ligatures and medicines to wounds. Five years later the examining function of the *protomédico* was given such importance that the universities of Castile were no longer able to award medical degrees to students until they had satisfied the *protomédicos*' (there were now three of these) examining board.[71] And at the king's request, set texts for this examination were specially written by the royal physician and *protomédico*, Luis Mercado. The king's commissioning letter specified that the texts were to be 'institutes', compilations of sound practice, which were to be printed and distributed throughout Castile[72]; and students who had completed university courses and two years' practical training had now to learn the texts by heart. Published in 1594, Mercado's institutes of medicine provided the digest of approved practice for physicians[73]; in the same year his institutes of surgery was published, containing sections on tumours, wounds and ulcers. The empirics also had a set book for their examination. For the *algebristas*, healers of fractures and dislocations, an illustrated manual was prepared, again by Mercado and at Philip's request. Based on the classical Hippocratic treatise, the text included chapters on the causes and signs of dislocations; methods of extending dislocated limbs; and the treatment for dislocations of the shoulder, wrist, hip, knee and spinal vertebra[74] (Plates 10a, b).

No set book was issued for apothecaries; but they too were the subject of the crown's attention at this time with the formulation

Plates 10a, b Illustrations from the set-book commissioned by the crown for the examination of bone-setters. They show prescribed techniques for treating dislocations of spinal vertebra: if the manual tugging did not succeed the second method was to be tried, applying gradual compression with the apparatus of beam and rigid supports. [*Instituciones que su magestad mando hazer al Doctor Mercado para el aprovechamiento y examen de los algebristas* (Madrid, 1599), Wellcome Institute library, London] of proposals similarly designed to establish a uniform, officially approved method of practice. In 1593 a royal pragmatic called for a joint commission of apothecaries and physicians to prepare an official pharmacopoeia for Castile; that was not realised. And around 1590 the king instructed his *protomédico* Francisco Valles to standardise the weights and liquid measures used by apothecaries in Castile. After consulting other physicians and the universities Valles published the officially approved set of units, making it clear in passing what the purpose was: 'it is desirable that the apothecaries of Castile use the Castilian mark because it is most convenient for government that in a republic there is no diversity of weights'.[75] The treatise also laid down approved techniques for preparing internal medicines by distillation – stills of copper, lead and tin were declared dangerous because metallic poisons entered the liquors prepared in such apparatus; glass alembics were to be used instead. The apothecaries had protested about this disruption of their routine, alleging that Valles' revised weights and measures were damaging to health[76]; nevertheless they were obliged henceforth to observe the new regulations.

## 3 *The king's medical establishment and the search for medicinal plants*

The *protomédicos* of Castile were also the king's personal physicians, though they were by no means the only practitioners to receive this honour. A large number of medical personnel attended the king, his family and the servants of the royal household. The sharp rise in their numbers in Philip's reign is striking, even after allowing for the long periods of absence from Spain of Charles V (his family who remained in Spain required medical servants). Under Charles the number of practitioners attending the royal household each year in the period 1530–55 had been 15–20. During Philip's reign this number rose to 24 in 1559, rarely fell below 30 in 1562–77, was never less than 40 in 1578–88 and reached a peak of no fewer than 47 in 1582.[77] Philip's sickly physique, constantly afflicted by gout, and the chronic illness of his son Don Carlos may go some way to explaining this considerable medical presence at court; but there is no full correlation with the health of the royal family – Don Carlos died in 1568 yet after this more and more royal physicians were engaged, and in 1589–98, the years of Philip's worsening invalidity, the number of royal physicians fell sharply to 20s and then to 10s.

Perhaps that final fall was due to economic considerations. Individually practitioners were not highly paid – the usual annual salary was 60 000 or 80 000 *maravedís*, about the same as a professor of medicine at Alcalá, and one half or one third of what a professor at Valladolid would be paid. It was no more than the chronicler Ambrosio de Morales received (80 000) and far less than the salaries of the most important royal officials, such as the president of the Council of the Indies – Luis Hurtado de Mendoza's salary was 500 000 *maravedís* in 1558. Nor did royal physicians' salaries increase with the rising prices; in 1589 Pedro López received the same salary of 80 000 *maravedís* he had first been given as long ago as 1547. Collectively the physicians' pay-roll was substantial: 857 000 *maravedís* in 1558; 2 372 166 *maravedís* in 1568; and 3 020 000 *maravedís* in 1588. Nor was this the total cost of medical attendance on the royal household, because resort was frequently made to additional practitioners who did not have the rank of *médico de cámara*, the coveted title held by physicians to the royal family. An example of this is the service given by Juan Fragoso. He was surgeon to Elisabeth de Valois, Philip's third wife, and after her death to Ana, the new queen; he also treated the king in the Escorial and at the royal palace of Valsaín in the forest of Segovia. By 1580 when he unsuccessfully appealed to be granted the prestigious title, he had served continuously for 20 years, and his salary for this is not included in the above-mentioned official pay-roll.[78] No wonder, with all of his other committments, Philip found it dificult at times to pay his physicians. In March 1582 after he authorised payments of 6000 ducats to Valles; 2000 ducats to each of two other *médicos de cámara*, Fernández de Vitoria and Diego de Olivares; 1000 ducats for Zamudio de Alfaro and another 1000 for Pedro Gálvez, the king approached the Council of the Indies to suggest where this money might be found. The reply was: 'at present the Council does not know, unless it comes from the sale of offices of the mints of Los Reyes (Lima) and Potosí, in which case they should be sold quickly'.[79]

How did these physicians reach this peak of the medical profession? For some the stepping-stone was a distinguished academic career. Fernando Mena, *médico de cámara* (1560–8), had been a professor at Alcalá in the 50s, then a leading centre for producing classical texts free of Arabic accretions; he had prepared translations with commentaries on Galen's texts on the pulse, urine and blood-letting; and had also written a widely-read work on the composition of medicines. Francisco Valles, *médico de cámara* (1572–92) and *protomédico* of Castile, had also come from Alcalá where he held a chair and published famous translations of Hippocratic texts with important commentaries which incorporated his own clinical observations. Luis Mercado had been a professor of medicine at Valladolid for 20 years when in 1592 he was appointed *médico de cámara*, a position he held until 1611. He had written numerous works defending Galen's doctrines and a treatise on the epidemic, a variety of typhus, which spread through Castile in 1570–1 in the wake of the transportation of the Granadine Moriscos. Others acquired a reputation through services to the clergy. Juan Gutiérrez de Santander who served Philip in the first ten years of the reign had previously been physician to the chapter at the cathedral of Sigüenza; and Pedro Gálvez (*médico de cámara*, 1554–90) had been physician to the Inquisition.[80] At least one of Philip's surgeons made a name for himself on the battlefield; Dionisio Daza Chacón, after graduating in medicine at Salamanca, gave surgical assistance to the wounded during Charles V's campaigns. Success in curing was the likely reason for the rise to royal service of unqualified empirics such as Francisco de Somovilla.

A few of the practitioners, like the surgeon César Barreta, were specifically appointed to treat poor servants of the royal household. Physicians who attended the royal family were expected to advise on healthy diet and inspect the meat and fruit which was to be consumed the next day; they were also assigned the traditional task of tasting drink before the monarch took his. They were forbidden to visit any patients outside who were suffering from smallpox, 'spotted fever' and other contagious diseases. Some of the practitioners were given narrowly specialised functions such as the treatment of hernia, dislocations and fractures; Francisco Martínez de Castrillo served for 20 years as Philip's dentist. There was usually little difference in the pay of all of these men; surgeons whether university-trained or not received much the same as the most distinguished physicians, and the same was true of the salaries given to empirics. The prestige of the royal appointment united them all.

The prescriptions written by court physicians were dispensed in the king's pharmacies. There were two in Madrid, one within the royal palace itself, and another outside for the king's servants. The servants' pharmacy, run by the apothecaries Juan and José de Arigón, was subject to an investigation in 1590 after complaints of neglect, slow service and short measures.[81] From 1594 the two pharmacies were amalgamated as the *botica real* supervised by Antonio del Espinar. A third royal pharmacy at the Escorial served the king and the monks. Several of its attractive polychrome vessels and caskets have survived,[82] decorated with lions rampant, the emblem of the Hieronymite monks; gridirons, symbol of the martyrdom of St. Laurence to whom Philip had dedicated the Escorial; and cartouches enclosing the names of the contained medicines.

The royal pharmacies were supplied with medicines prepared by royal distillers using plants grown in royal gardens. Throughout his reign Philip displayed a pronounced interest in the acquisition and cultivation of medicinal plants. in 1557–8 he had approved a draft contract with Francisco de Mendoza, son of a former viceroy of New Spain, to cultivate various plants of medicinal importance, growing them in the Peninsula or in New Spain. They included china-root (used for gout), sandalwood, and the medicinal spices: ginger (a carminative), cloves, pepper and cinnamon. Philip agreed to donate land for this purpose, granting Mendoza a monopoly on sales and sharing the profit with him.[83] No record has been found of the realisation of this business enterprise; but Mendoza is usually credited with introducing ginger into New Spain from the East. Elsewhere cultivation was fostered by the king solely for the health of his household. In the lush grounds which he helped to plan at the palace of Aranjuez the king gave space to medicinal plants. The gardens here were attended by Holbeeck, a Fleming, who from September 1564 also served as 'royal distiller of essences and liquors', chemically extracting medicinal substances by distilling infusions of plants. In 1566 after preparing various essences in this way for the royal pharmacy in Madrid, he asked for silver flasks of one *azumbre* capacity (about 2 litres), to be manufactured for their transport 'because the preparations are laborious and expensive and it is undesirable to risk sending them in glass vessels';[84] this was done. And at the same time the king employed Luis de León, a physician skilled in the knowledge of simples (medicinal herbs) to travel through Castile in search of them; local justices were requested to supply the doctor with mules to carry the herbs and plants collected, and provide suitable accommodation with facilities for 'storing them with care'.[85] The plants were to be brought to Madrid. The royal pharmacy in Madrid was later to be assisted by the appointment of another distiller, Giovanni Vincenzio Forte of Naples; his alembics and procedure were observed with approval by the king's minister Cardinal Granvelle, who reported that 'he will give good service especially when the garden of simples will be finished'.[86] At the Escorial another physic garden was planted and supplied the raw materials for the several distillers on the site who with elaborate apparatus prepared medicinal liquors for the monastery's pharmacy (q.v. p. 14).

More important than any of this in the long term was Philip's decision to send his *médico de cámara* Francisco Hernández across the Atlantic in seach of new medicinal plants. The expedition produced the greatest scientific achievement of the reign. The Indies had already acquired the reputation of a source of marvellous panaceas. Around the time of Philip's accession Las Casas was extolling the medicinal virtues of the balsams provided by the liquidambar and 'a great multitude' of other aromatic trees of the New World. And his argument for the superiority of Hispaniola to every other isle in the world was based on its 'infinity' of health-giving trees, above all the guaiacum whose bark was an assured remedy for syphilis and all diseases originating from 'the cold and the moist'. These were treasures greater than silver or gold.[87] The same contrast of New World medicines and precious metals was made by Nicolás Monardes in the preface to his widely-diffused *Historia Medicinal* (1565–74). This physician of Seville acquired American plants, shipped to this centre of the Indies trade, and grew them in his garden; his book familiarised Europeans with balsams, sarsaparilla and tobacco, then valued principally as a panacea. And now Philip was planning a scientific expedition to investigate these exotic plants more fully than before. What was the king's interest? Commercial motives cannot be ruled out, particularly in view of Hernández's letters after arrival which promised Philip that 'these Indies would supply medicines to the entire world'.[88] But the chief aim may well have been to promote the health of his subjects, keeping them fit for the tasks of peace and war, at a time when conventional medicine offered little protection from disease.

Hernández, a medical graduate of Alcalá, had practised in Toledo and Seville, during which time he also made botanical excursions within the Peninsula. He was appointed to Philip's court in 1568. In January 1570 he received the king's instructions to depart for New Spain to collect all the information he could on that land's medicinal plants by consulting local 'physicians, surgeons, herbalists, Indians and others who appear to you knowledgeable on these matters', thereby learning about 'the virtues of the said medicines, the places where they grow and whether there are different species of them'.[89] Hernández was asked to perform experiments and make observations to test this information and to record all facts confirmed. Extraordinary medicines, herbs and seeds were to be sent to Castile. And when this had been completed he was to proceed to Peru to do the same there. He was given five years for the mission and graced with the grandiose title of 'our *protomédico general* of our Indies, isles and *tierra firme*'.[90] At the same time Philip sent instructions to his viceroys to assist the expedition by providing experts to gather herbs and artists to draw them.

Hernández sailed from Seville in September 1570 accompanied by Francisco Domínguez, the cosmographer, sent with him to prepare a geographical description. The ship called at Grand Canary, Hispaniola and Cuba; on each island Hernández studied the flora. He landed at Vera Cruz in February and soon set off by mule on extensive travels with artists, experienced plant-hunters and an interpreter to question Indian medicine-men. By March 1573 he wrote to Philip that he had completed four illustrated volumes containing descriptions of over one thousand plants. They had never been seen in Europe and possessed 'great virtues of unbelievable and enormous benefit'. He was entirely justified in feeling that he had surpassed Dioscorides; the Ancients had known nothing of this. And in another letter which returned to the theme of excelling beyond the achievements of classical antiquity he assured Philip that this expedition would bring the monarch much greater fame than Alexander received from his patronage of Aristotle's research in natural history.

He covered large areas of the huge territory of New Spain, and at several hospitals experimented on the sick with medicines from the plants collected. But his health had suffered and he was in no state to contemplate the completion of his mission with travels in Peru. The five years' time-limit had also elapsed, and Philip granted a year's extension. Hernández sent the king several volumes with descriptions of plants, their forms, virtues, and the mode of application of medicines prepared from them. The manuscripts went to the Escorial's library, some of the illustrations were used to decorate Philip's living rooms. Hernández returned to Spain in 1577 with more volumes, bundles of illustrations, sacks of seeds, and tubs with live plants – some, planted in the gardens of the alcázar in Seville, survive still as large trees.

Instead of publishing a multi-volume work, Philip, much to Hernández's disgust, decided on an abridgement which physicians could use as a manual. The task was assigned to Nardo Recchi, a Neapolitan physician. Juan de Herrera, Philip's architect, arranged for sample drawings (see plate 11) to be made of Hernández's original illustrations. Estimating that a total of 400 engravings would be needed at a cost of 1500 ducats, he commented that, in view of the great benefit this work would bring to all 'I know of nothing on which such a sum could be better spent'.[91] Nothing was however published until the seventeenth century when texts based on Recchi's epitome appeared in Mexico and Rome. Hernández's manuscripts were destroyed by a fire in the Escorial in 1671; but fortunately other drafts survived for publication in later centuries. Hernández's work describing thousands of species of the flora of Mexico has never been superseded.[92]

Nothing comparable was done for the flora of Peru, but the *relaciones* from that region gave strong indications of a rich source of materia medica. The *corregidor* of one locality in the viceroyalty reported that he had witnessed the healing of wounds and contusions by the application of herbs used by the Indians, of which there were so many 'that it would take over two reams of paper to list them'.[93] This was the land which in the next century would provide the most valuable of all American medicinal plants: the cinchona tree, whose bark contained quinine. Not all of the Indians' herbs were welcomed by the crown. In 1572 Viceroy Toledo prohibited the use of the juice of a species of yucca because it was 'said to be poisonous and causes illness'; it had been a drink of the Andean Indians.[94] And in 1579 it was reported from one locality of New Spain that some local herbs used by pregnant women for abortions had been banned because they were dangerous.[95]

Philip's involvement with medicinal plants continued into his final years. He wrote to Mathias de Albuquerque, his Portuguese

Plate 11 Engraving of a medicinal plant from the New World; it was made as a sample illustration for the abridged edition, commissioned by the crown, of Francisco Hernández's voluminous survey of the flora of New Spain. (IVDJ 99/190; 1582).

viceroy of India, for information on oriental herbs, and in reply received descriptions of 150 plants with details of how to prepare medicines from them and the diseases for which they served as remedies.[96] And he expressed interest in acquiring a botanic garden in Seville after the death of its owner, the physician Simón Tovar. This was a well-known garden from which Tovar sent seeds and information on acclimatisation to numerous European correspondents including the distinguished botanist of the Netherlands, Charles de l'Escluse. The king was told that the garden might be bought for 1600 ducats and that 'it would be one of the outstanding things in Spain; indeed it might be said that it is already so, since Italians and other foreigners have come here just to see this'.[97] There is no associated document recording the sale of the garden to the king. Another garden had been purchased in Madrid by the king from Diego de Burgos, royal apothecary, with a view to incorporating it with the adjoining gardens of the royal palace. This was now developed apparently for the supply of the king's pharmacy. Honorato Pomar, professor of medicine at Valencia, was appointed royal simplist to cultivate all types of herbs and medicinal plants which could be found. On Pomar's advice this garden was preferred to a site in the Casa del Campo, partly because it would be more convenient for Philip's *médicos de cámara* to visit and acquaint themselves with unfamiliar herbs.[98]

## 4 *Medical services for the military*

Of all his subjects Philip could least afford to neglect the health of his armed forces. If Spain was to be triumphant her armies and fleets must always be in a fit state to fight; the spread of disease was capable of delivering a far more crushing blow than any the Turk might inflict. If the outbreaks of contagion among the civilian population could be left to the municipalities to deal with, the crown could not stand aside when they threatened the health of its soldiers and seamen, and emergency measures were taken at the first signs of danger.

The crown's priorities were apparent in the preparations for the annexation of Portugal. In January 1580, while a large armada was being assembled in the bay of Cádiz, it was learned that many of the infantry arriving there from Sicily had died of a sickness. Philip instructed the marquis de Santa Cruz, commander of the armada, not to keep these sick arrivals in Cádiz or any other port on that coast, but to send them inland to the towns of Jerez, Jimena and Vejer; this would avoid contagion of the air in the vicinity of the fleet and 'the consequent irreparable harm to the armada'.[99] In the same month the king took further precautions to protect the armada from the pestilence in Ceuta; all contact with ships from the Barbary coast was to be avoided.[100] And when later that autumn, soon after the successful invasion, Lisbon was in the grip of an epidemic of '*catarro*' (perhaps influenza with pneumonic complications) which eventually killed 6000, Philip ordered the evacuation of houses close to the city's castle of São Jorge to prevent the infection reaching the 1200 soldiers garrisoned there; and he asked for the cavalry to be stationed in 'as healthy a place as possible'.[101] Spanish infantry were also quartered in houses on the outskirts of the city. Philip ordered their speedy removal so that the owners, who had been displaced to the centre of the city with consequent crowding and greater risks of the spread of infection, might return to their homes in the interest of the health of both the citizens and the occupying army.[102]

The crown's fears of an epidemic among the troops may have been played upon by host civilian populations resentful of the billeting imposed on them. That at any rate was suggested in January 1584 when, amidst the alarm of officials of the armada stationed in the bay of Cádiz at the news of pestilence in neighbouring Jerez and talk of preparations for evacuation, one observer said it was nothing but a ruse to get rid of the soldiers.[103] Tensions between civilians and the military were also generated by the return of expeditionary forces, invariably carrying large numbers of sick into Peninsular ports. This occurred in September 1583 with the return to Cádiz of the force from the Azores, where the isle of Terceira had been captured from the supporters of the Portuguese pretender Dom Antonio and his French allies, thereby securing a strategically important port of call for the king's silver fleets. But the victorious soldiers were not welcomed in Cádiz as more than 2000 sick disembarked, some of them captured French and Portuguese. The sick were taken to the city hospital, but there were too many for all to be treated there; additional hospitals had to be specially formed and some of the sick were put in houses. The city complained to the secretary of war that the soldiers had brought a contagious disease, *tabardillo* (a form of typhus) which had spread killing one seventh of the citizens. Evidence was taken from city surgeons and physicians who testified that Cádiz was a healthy place before the troops arrived. The city appealed for the removal of the military sick fearing that otherwise a worse pestilence would erupt on the few residents who had not yet fled.[104]

The flight of the civilians of Cádiz was a common response to contagion; for outbreaks of plague it was even advised by physicians as a remedy. But soldiers were not free to do this; that would have been desertion. They were evacuated on their commanders' orders but strategic sites could not be readily abandoned. In April 1590 pestilence caused the inhabitants of Colivre (Collioure) to flee. This coastal town near the French border was in a fortified zone centred on Perpiñán. A skeleton force was left at Colivre and in neighbouring fortresses to defend them from Moorish galleys and the French. And if the disease reached Perpiñán, four leagues away, plans were ready for enclosing the soldiers within the castle.[105]

The scant resources of civil hospitals were frequently strained to the limit by military demands. The mere presence of billeted soldiers could affect the functioning of a local hospital. The administrator of one royal hospital explained why it was undesirable to billet infantry in his locality. His hospital was funded by the sales-tax levied in 12 neighbouring *pueblos*. But the inhabitants were poor and their income, derived from supplying firewood and charcoal, had been badly affected by drought; if they were required to provide food, wine and firewood for soldiers billeted on them, even less of their money would find its way to the hospital. The Council of War was persuaded by this to quarter the soldiers elsewhere.[106] The much more common complaint from the hospitals was that they could not cope with the numbers of sick military. The town of San Lúcar de Barrameda, at the mouth of the Guadalquivir, had just one hospital, originally established for the treatment of the sick poor and maintained by local alms. But now the town had become a busy point of departure and arrival of armadas of the Indies, and many sick and wounded soldiers and seamen were constantly sent to the hospital for treatment. The hospital could not afford to supply the necessary physicians, pharmacy, beds and food and therefore approached the king for financial assistance.[107]

Civil hospitals on the north coast were overwhelmed in the autumn of 1588 by the return of the survivors of the Invincible

Armada sent against England. There had been resort to civilian medical resources even before that expedition had reached the Channel. Three weeks after departing from Lisbon the Armada had been struck by a severe storm, causing it to scatter. As the damaged ships re-grouped in La Coruña, there was found to be fever and much sickness from food which had gone rotten. On the king's orders the archbishop of Santiago arranged for medical assistance, providing money for the purchase of medicines – a Dr Rubio and *licenciado* Zárate, a surgeon, were sent to Santiago to buy them – and organising the transfer of 389 sick to the hospital in La Coruña and of 43 others with contagious diseases, for whom there were no facilities there, to a hospital in Santiago. The archbishop received progress reports from *licenciado* Martínez, the physician attached to the hospital at La Coruña, who was treating these men of the Armada.[108] By the end of July the Armada was ready to set out again; two months later its battered vessels limped home with men sick, dying and starving. The hospital at San Sebastián was unable to cope when Miguel de Oquendo, commander of the Guipúzcoan squadron, landed at the nearby port of Pasajes with seventeen companies of infantry. He asked the justice of San Sebastián for houses to form a hospital for the sick; these were provided. Beds were taken from residents under promise of payment – the owners were still pressing for their money five years later; the beds (and houses) had been destroyed within a few days by a blaze at the improvised hospital caused by fires which had been lit for the sick. Local apothecaries supplied medicines to the value of 315 917 *maravedís*; half of this was still unpaid a year later.[109] At Santander over 2000 sick disembarked; its poorly endowed hospital could not manage and additional emergency hospitals had to be set up. The nearest concentration of physicians was over 30 leagues away at Valladolid. Philip wrote to the municipal authorities asking them to send medical practitioners to Santander with as much medicine as possible. Valladolid dispatched a physician, a surgeon, two apothecaries and 50 mules loaded with medicines and supplies.[110] More than 60 of the Invincible's sick were sent to another university medical centre, Salamanca. They were treated at the city's general hospital by Dr Espinosa, professor of medicine and natural philosophy, and salaried physician of the hospital. Most recovered, but about ten died from a disease which was alleged to be contagious, because students, medical staff and townspeople who gave voluntary assistance were all infected; one of the physicians died. Espinosa ordered a surgeon to perform post mortem examinations of the dead soldiers 'to see if I was erring in my treatment'. From the decayed condition of the internal organs and the presence of putrid matter he concluded that his diagnosis had been correct and that the cause of disease was rotten victuals and impure drinking water. He expressed great relief that these infectious invalids had not arrived a few months earlier in the heat of the summer because that would have been the beginning of 'the cruellest of pestilences'.[111]

The use by the military of civilian health facilities was supplementary to other services which existed in a well-developed form within the armies and fleets. In the provision of health care for her forces Spain did far more than any other European state. At its lowest level this took the form of the attachment of a practitioner to garrisons, military units and ships. At the beginning of his reign Philip had introduced regular medical services on galleys through his instructions to the captain-general of the galleys of Spain to employ 'one physician with three or four barber-surgeons' and to keep on board 'a pharmacy stocked with good medicines'. That was for the entire squadron; individual galleys had to make do with the rudimentary skills of 'a Moorish or convict barber'.[112] On land soldiers stationed near the border with France were regularly attended by physicians. In the 1580s Dr Quesada, a Navarrese physician, was caring for soldiers at Fuenterrabía and San Sebastián. And in Navarre the infantry and their families were treated by Dr Guevara and *licenciado* Bayona, who were paid as enlisted soldiers within a particular company, receiving four ducats a month for this; they were given an additional five ducats a month for attending a hospital for the infantry.[113] In Italy every unit of Spanish troops had its physician and surgeon since the time of Charles V. And the *tercios* (units of about 2500 men) of Philip's Army of Flanders were each provided with a surgeon-major.

But the most impressive organisation of Spanish medical care for the military was in the creation of special hospitals. Of these there were three different types: permanent foundations; temporary base hospitals fixed in specific locations; and field hospitals, temporary and mobile.

Pamplona seems to have been the only permanent military hospital in the interior of the Peninsula, founded sometime after 1574 when the Council of War recommended the acquisition of a site and a donation of 1000 ducats.[114] The maintenance of this and all other Spanish military hospitals depended on grants from the crown and smaller sums collected by deducting one *real* a month from every soldier's pay. Unfortunately both sources of income frequently dried up because of the crown's financial crises; soldiers were left unpaid for long periods with consequent loss of the one *real* contribution. The hospital administrators begged the crown for money, complaining that because the infantry had not been paid for three years the hospital was in no position to admit even a few sick.[115] The same failure of resources led to disastrous consequences at the hospital which was being established on the coast at Cartagena around 1584. Don Juan of Austria, Philip's captain-general of the sea, and Luis de Requesens, lieutenant-general of the galleys of Spain, had informed the Council of War of the need for a special hospital at this port, where the sick of assembling armadas could be treated effectively and at less cost to the crown. The municipality had agreed to the erection of an infirmary in the main street, and an administrator had been appointed; 12 lay friars provided the nursing, and there was a chaplain. But no endowment was given, and hundreds of sick soldiers disembarking from galleys in 1584 and sent to the hospital found there was no food, medicines or beds for them; 775 of them were said to have perished. When the king asked what was needed to complete the foundation of this hospital he was told that he would have to provide 2000 ducats a year, and a further 4000 ducats for moving the hospital to a healthier site on the outskirts of the city.[116] At Gibraltar the king provided a site above the arsenal for a hospital to treat the crews of galleys, but the purchase of a house belonging to the count del Castellar gave a more spacious building. Here the king was told of the importance of offering adequate salaries to attract surgeons, who were in short supply; they were said to be indispensable not only in time of war but constantly because without their treatment 'many galley convicts become useless through sores and other disorders'.[117] A permanent hospital for soldiers functioned at Perpiñán and another was being planned in 1597 at Santander for the use of armadas.

In the Spanish Netherlands the Army of Flanders was provided with an elaborate system of health care based on the military hospital established in 1585 at Mechelen with a team of physicians, surgeons and student-surgeons; nursing was done by religious brethren sent from Spain, members of the hospital Order of San Juan de Dios.[118] Soldiers of the twin African fortresses of Oran and Mers-el-Kebir were treated at the military hospital of San Bernardino. There were far fewer troops here than in the Netherlands and the medical staff consisted of just one physician, one surgeon, an apothecary and his assistant, and two nurses. Their combined salaries, along with that of the administrator and chaplain came to 300 000 *maravedís*, which the hospital was barely able to meet, leaving little for the treatment of the sick. Its expected annual income, rarely collected in full, was 350 000 *maravedís*, mostly from the levy on soldiers' pay (amounting here to 5000 *reales* or 152 000 *maravedís*), the rest coming from rented accommodation in Oran and an annuity. When the king suspended soldiers' pay, the physician, Dr Pedro Crietes, was deprived of his salary. The crown's response to repeated appeals for assistance was to donate occasional grants of money, some of which were proceeds from the wheat and meat given to the crown by subject Moors in return for permission to sow crops and graze cattle on land adjoining the fortresses. The administrator found himself constrained by the hospital's statutes which prohibited the reception of patients suffering from syphilis or other contagious disease, yet such cases were frequent at Oran. He decided to admit them, doing what he could to keep them apart; the reported success of the treatment persuaded the Council of War to recommend the allocation of 1000 ducats for the creation of an isolation ward.[119] Another African fortress, Peñón de Vélez, also had its military hospital with nursing by brethren of the confraternity of La Santa Vera Cruz. The apothecary there, Cristóbal de Aguilar de Viana, was also a serving soldier; when he travelled to Málaga with a warrant for money due to him, he was told that there was nothing with which to pay him.[120]

The most distinctive feature of Spain's medical provision for the military was the use of temporary hospitals, set up to meet immediate needs and then dissolved as soon as they were no longer required; their administrators, invariably clergy, then returned to their churches, and their medical practitioners, always in demand, moved on to wherever else they were needed. Two such hospitals were formed in Lisbon in the autumn of 1587 while the Invincible Armada was being prepared there. They were needed because the slow preparations had kept men for long periods on board ship with insanitary conditions and poor food. In October there were as many as 500 sick in these hospitals; a month later the number had risen to 700 and more were lodged in the city hospital and in private houses.[121] After the Invincible set sail in May these hospitals ceased to exist, and sick seamen in Lisbon either had to go to the pressed city hospital – it served for civilians in a wide area around the city – or more often, so it was alleged, 'to places so indecent that they risk losing their souls as well as their bodies'.[122]

The Armada carried with it at least seven physicians and five surgeons; some of them were never to return.[123] They constituted a hospital afloat with an administrative staff. In addition other practitioners were attached to particular units.[124] When the Armada repaired at La Coruña after the June tempest medical certificates were issued by physicians on board declaring some soldiers unfit to serve; they were then discharged with permits to leave by the commander, Medina Sidonia.[125] And when the battered fleet returned, El Ferrol became the centre for its reconstruction, and in October 1589 a temporary hospital had been established. Apart from caring for the sick stationed here it also served as a supply base for a hospital formed in May 1591 at Blavet in Brittany for Philip's invasions into France, abortive efforts seeking to keep Henry IV from the throne. Flyboats took medicines and diets for the sick from El Ferrol to Blavet, amidst complaints that supplies were insufficient, stale or stolen.[126]

The ships also took physicians and surgeons from El Ferrol leaving the base short of practitioners. At first there had been a total hospital staff of 53 which included 13 nursing brethren, 3 barbers, 3 surgeons, 3 physicians and an apothecary.[127] Some of the qualified practitioners who went to Brittany were difficult to replace. The hospital was dissolved in December 1591, re-established again, and finally dissolved at the end of 1593. Its first administrator, Dr Manso, canon of Burgos, had meanwhile proceeded to Aragón where Philip had sent 12 000 soldiers to quell the uprising in Zaragoza; soon he was organising a military hospital at Jaca; others were set up in mountainous terrain at Monzón and Barbastro.[128]

Field hospitals, used by the Catholic Monarchs during the reconquest of Granada, were frequently organised for Philip's armies. In 1559 a 'portable' hospital was designed by a clergyman for the expedition sent to recapture Tripoli. Its rectangular plan, sent to the king, shows provision for 200 patients with adjoining rooms for physicians and clergy, a dispensary, kitchen and place of prayer; and, of dubious advantage, its inventor said it would also serve to store munitions. Judging by the praise given to it by Medinaceli, the expedition's commander, during his preparations, there is reason to suppose it was actually used.[129] A field hospital was set up during the Morisco uprising at Los Padules, on the edge of the Alpujarras, the centre of the action; in May 1570 lists show that 62 sick or wounded had been received, another 30 had been sent on to the city of Guadix, some ten leagues away, perhaps to a civilian hospital there.[130] In April 1580 Gaspar de Mendoza, clergyman, was on his way to Seville with 3000 ducats to buy medicines and beds for the field hospital he had just been appointed to administer for the invasion of Portugal. Formed under tents in Extremadura where Philip's army was concentrated, its small team of practitioners divided when the fighting began, some of them accompanying the advancing troops, the rest remaining at the base.[131] And in the continuation of this war in the Azores, the sick and wounded of Philip's expeditionary force were treated in field hospitals at Angra on the isle of Terceira.[132] Throughout the long campaigning in the Netherlands, mobile field hospitals were in operation for the Army of Flanders, supported from the fixed hospital centre at Mechelen.

The nomination of military physicians and surgeons was sometimes made by military commanders; often the advice of the king's *protomédico*, Valles, was sought. When a military hospital was created for the garrison defending the Canary Islands, no physician could be found on Tenerife and after consulting Valles, Antonio Mantilla was brought from the Peninsula.[133] In the same way that the crown approached archbishops for the nomination of clergy with the requisite Christian virtues to serve as hospital administrators, Valles was consulted by the Council of War on the appointment of military physicians; he was sent a list of applicants and asked to comment on their merits and suggest salaries. In his replies he said Juan de Rubio was 'a good doctor, very suitable to serve' in the Invincible Armada; but that Dr Duarte Berdugo, who had applied to attend the army in Portugal, was 'mediocre and should be appointed only if others could not be found'.[134]

Elsewhere the appointment of practitioners was controlled by military *protomédicos* who were given analogous powers to their civilian counterparts. Pérez de Herrera, physician of the galleys of

Spain, was promoted to *protomédico* of the same galleys with authority 'to examine all persons wishing to practise as physicians, surgeons or apothecaries to these galleys, and finding them able you may issue them with examination certificates signed by you'; he also supervised the pharmacies kept on the galleys.[135] Earlier in the reign a similar position was filled in Philip's duchy of Milan with the appointment of Lope Bastardo as *protofísico galeno* to the Spanish army there. His duties were to engage adequately trained practitioners and to dismiss any found wanting; none could practise without his approval. He also guaranteed the quality of medicines dispensed to the troops. All ranks were ordered to hold him in respect and obey him 'in all that is necessary'.[136]

The lives of military physicians were no more at risk than a city doctor who attended the victims of pestilence. But there were other dangers and it is no surprise to find them suffering the same misfortunes as soldiers. Rodrigo de Molina was a physician in the fortress of Oran when he was ordered to go with the soldiers on the expedition (1558) to Mostaganem, the port 12 leagues distant, through which supplies were coming for Turkish operations in the western Mediterranean. The Spanish force suffered a crushing defeat, and Molina, one of 12 000 who were captured, had to be ransomed.[137] Pérez de Herrera was wounded by an arquebus during the assault on Terceira. And at least two physicians died on the Invincible Armada.

Low pay – even the prospect of not being paid at all, like the soldiers – was probably a greater deterrent to their recruitment than the dangers of war. The initial offers by the crown were frequently refused, and there is evidence of subsequent haggling to secure their indispensable services. But money was not the only reward; there was also the lure of a prestigious title. Alonso de la Cerda who attended galleys stationed in Lisbon said he did this 'for love of God and without any pay', but he asked to be given the title of surgeon-major of those galleys.[138] Whatever the attractions, some practitioners followed the military from campaign to campaign. Antonio Pérez served as surgeon-major in the defeat of the Morisco uprising, and later in Flanders, the Levant, Portugal and the Azores. Daza Chacón was the most experienced of all. He had been a military surgeon since 1543 when he tended the wounded of Charles V's army at the field hospital in Valenciennes; subsequently he served in Granada, Lepanto and Portugal. Even when they remained within the Peninsula, the work of these practitioners could be hampered by language difficulties – the numerous Basque soldiers and seamen were often unable to speak Castilian. During the preparation of the Invincible in Lisbon there were over 1000 sick Basque recruits; they described their symptoms to a Basque notary of the Armada hospital and through the assistance of this interpreter the Castilian physicians said many lives had been saved.[139]

Medicines were needed by soldiers and seamen for the wounds of battle, for syphilis and the fevers contracted from unhygenic living conditions, unwholesome food and impure water. An offer to provide such medicines in a more effective form by alchemical distillation interested the king sufficiently to bring Leonardo Fioravanti from Italy to prepare some in Madrid. This university-trained physician had approached the crown with proposals to improve the treatment given to the king's troops in Italy, which he alleged was of a low standard.[140] He attributed all disease to poisons. Peccant humours were produced in the stomach by the corruption of food; and, in the casualities of war, poisoned wounds were caused by iron projectiles. He was confident that all could be cured with the liquors which he distilled from a variety of substances, the poisons being expelled by an occult virtue in the quintessences so prepared.

There is no record that any of these Paracelsian remedies were ever administered to Philip's forces. The documents show instead a reliance on traditional medicines, predominantly herbal but also some conventional mineral and animal substances; there are also notable examples of newer medicines, but these were not alchemical preparations. A list of the stock needed by the pharmacy of the armada at El Ferrol to compose medicines in the course of a year[141] shows the use of numerous well-established plant medicines: infusion of roses (12 *arrobas*); colocynth (6 *libras*); myrrh (12 *libras*); mastic (1 *arroba*); cardamoms (2 *libras*); cloves (4 *libras*). Mineral substances included mercury (20 *libras*), one of the remedies for syphilis; and gold (1000 leaves). Substances of animal origin included the bezoar stone (half a *libra*), a much-esteemed antidote for poisons. The pharmacy consumed large quantities of honey (50 *arrobas*) and sugar (100 *arrobas*), sweeteners for electuaries. Newer plant medicines of American origin are also listed: china-root (1 *arroba*); mechoacan (8 *libras*) and tacamahac (20 *libras*).

These, like many of the medicines of the time, had no single specific application but were thought to be valuable for a range of disorders, some of them approaching panaceas.

Medicinal plant substances imported from the Indies were also sent from this pharmacy in El Ferrol to the military hospital in Brittany. Records of these show that in 1591 consignments included china-root (6 *libras*); sarsaparilla (50 *libras*); guaiacum (75 *libras*); and – by far the largest recorded quantity of any medicine sent to Brittany – cassia fistula (6 *arrobas*).[142]

These lists show that soldiers' wounds were being treated with soothing dressings[143]: the 'oil of Aparicio', the herbal composition discovered by the empiric Aparicio de Zubia and which he had used for arquebus wounds and powder burns on the battlefield of Saint Quentin. The older drastic treatment, involving the application of boiling oil to gunshot wounds, had been rejected in the previous reign by Daza Chacón. He also saved Spanish soldiers the suffering inflicted by the technique of cauterizing limbs after amputation to stop haemorrhage; he applied ligatures instead. Neither of these was an original contribution; they had both been introduced by the French military surgeon Ambroise Paré.

Medicines were expensive and constituted a large fraction of the cost of running a military health service. There are surviving records of the expenditure because the king wanted to know exactly how much it was costing him to maintain the armada hospital at El Ferrol. Its administrator, Dr Manso, prepared a detailed account for a year to allow for the monthly variations in prices and in the numbers of sick admitted. In October 1589 – November 1590 the cost was 343 849 *reales*: the largest items of expenditure were food and wine (77 799); staff payments (76 960); and medicines (57 887).[144] This statement went to the Council of War where it was well received.[145] But on other occasions there had been concern over hospital costs. In Lisbon in 1581 a royal *veedor* (inspector) of the armada reported that 'half of the expenditure on military hospitals is badly spent' and advised using civilian hospitals instead, offering them the levy on soldiers' pay (a small part of what would have been needed).[146]

When the crown could not find the money for medicines they were taken on credit. The purveyor of the armada at Cartagena advised the king to re-pay the 8208 *reales*' worth of medicines supplied on credit two years before by the apothecary Juan Frances for the hospital at Oran; otherwise it would make it more difficult to find credit in the future. Juan de la Rubia, another apothecary of Cartagena, had to wait even longer; he was still pressing the Council of War 12 years after he had given medicines on credit, to the value of 9570 *reales*, to the galleys of Spain.[147] The reduction of royal funds to military hospitals is especially noticeable during the 1590s when Philip had committed Spain's resources to campaigns in France, a new burden to add to the unending war in Flanders. From Perpiñán came desperate appeals for money for the hospital lest the patients die of starvation. At El Ferrol medicine and food had to be taken on credit. At Jaca the army hospital owed local apothecaries, physicians and other civilians 5000 ducats and 'now we can find no one to give credit'.[148] These were the effects on medical care of yet another financial crisis which culminated in the 'bankruptcy' of 1596. Medical care, like much of the technological enterprise of Philip's reign, was stimulated by the succession of wars and hindered by their exhaustion of the royal treasury.

## *Notes*

1. AGS:GA 174/35, royal *cédula*, 9 March 1561, Toledo.
2. Diego de Encinas, *Cedulario indiano*, vol. 1 (Madrid, 1596), p. 219.
3. *Ibid., ibid.*, pp. 219–20, royal *cédulas* of Prince Philip and the princess regent, 18 May 1553, Madrid and 12 September 1556.
4. *Colección de documentos para la historia de la formación social de Hispanoamérica 1493–1810*, ed. R. Konetzke, vol. 1 (Madrid, 1953), pp. 444–5; the king to Martín Enríquez de Almansa, 2 June 1569.
5. J. Muriel, *Hospitales de la Nueva España*, vol. 1 (Mexico, 1956), pp. 203 and 210 f.
6. L. Martz, *Poverty and Welfare in Habsburg Spain: The Example of Toledo* (Cambridge, 1983), pp. 54–7.
7. *Ibid.*, p. 61 f. tells the story well. The same work describes Philip's new poor laws of 1565 which sought to distinguish the deserving poor from vagabonds, and included measures for confining persons with contagious diseases, p. 29 f.
8. T. González, *Noticia histórica documentada de los célebres minas de Guadalcanal desde su descubrimiento en el año de 1555, hasta que dejaron de labrarse por cuenta de la real hacienda*, 2 vols., (Madrid, 1831), vol. 1, pp. 163, 519, 579; vol. 2, pp. 352 and 355.
9. AGS:DC 46/66, 'Condiciones que hizo el virrey con los mineros por quatro años que fueron del año de 77 a 81'.
10. Archivo Histórico Nacional, Madrid: *Ordenes Militares (Archivo de Toledo), expediente* 37,888, unfoliated, *pleito civil* brought by Leonardo Salto, *alcalde* of the convict prison at Almadén, against *licenciado* Lupercio de Viveros, physician, December 1589.
11. Luis Capoche, *Relación general del asiento y Villa Imperial de Potosí* [1585], ed. L. Hanke (Madrid, 1959), p. 175.

12. *Gobernantes del Perú*, ed. R. Levillier, vol. 8 (Madrid, 1924), p. 14; *ordenanzas* of Viceroy Toledo, 3 October 1572, Cuzco.
13. *Recopilación de leyes de los reynos de las Indias* (Madrid, 1681), *lib.* vi, *tít.* xiv, *ley* ii, 18 October 1569. Madrid.
14. *Ibid., lib.* vi, *tít.* xiv, *ley* ii, 11 June 1573, Madrid.
15. *Gobernantes del Perú*, vol. 8, pp. 28 and 30. The viceroy also tried to prevent Indians selling maize, given to them as food, in order to buy coca leaves; the drug removed the pangs of hunger, but he said Indians were becoming ill through not eating.
16. The constitution is reproduced in J. Zarco, 'El hospital de El Escorial', *La Ciudad de Dios*, **132** (1923), 415–22; **133** (1923), 5–13 and 100–4.
17. Examples of contracts between town and physician are given in L.S. Granjel, *La medicina española renacentista* (Salamanca, 1980), pp. 68–9.
18. J.M. López Piñero, 'The medical profession in 16th century Spain', *The Town and State Physician in Europe from the Middle Ages to the Enlightenment*, ed. A.W. Russell (Wolfenbüttel, 1981), pp. 85–98.
19. AGS:GA 126/244, city of Seville to the king, May 1582; *ibid.*, 126/255, conde de Villar, *asistente*, and other officials of the city of Seville to the king, May 1582.
20. *Ibid.*, 138/175, *consulta, Consejo de las Indias*, 1582.
21. *Ibid.*, 140/46, advice of Doctors Valles and Fernández de Vitoria on the pestilence in Seville, 1582.
22. B. Bennassar, *Recherches sur les grandes épidémies dans le nord de l'Espagne à la fin du XVIe siècle* (Paris, 1969), pp. 52 and 75–6. More information on the action taken by the crown and the towns during this outbreak is contained in B. Vincent, 'La peste atlantica de 1596–1602', *Asclepio*, **28** (1976), 5–25. This shows that some of the affected towns were too poor to cope with the plague: the small town of Vera, near Almería, could not afford the services of an apothecary; it also indicates the crown's role in overcoming the disruption of commerce: cities were persuaded to restore commercial links with Málaga in 1602, when the plague had already retreated, only after Philip III had sent a *cédula* requesting this to all the *corregidores* of Castile.
23. Luis Mercado, *Libro en que se trata con claridad la naturaleza, causas, providencia, y verdadera orden y modo de curar la enfermedad vulgar, y peste que en estos años se ha divulgado por toda España* (Madrid, 1599). The preface contains Philip III's letter to Mercado, 14 July 1599, commissioning him to prepare a Castilian edition; and on p.23 Mercado states that the original Latin version was written by order of Philip II. Disagreement amongst the physicians on how to treat the disease led the crown to arrange a conference: in 1599 eight physicians and three surgeons considered whether it was best to cauterize carbuncles in the hope of a rapid cure, or – this was preferred – to proceed with more gentle surgery; Andrés Zamudio de Alfaro, *Orden para la cura y preservación de las secas y carbuncos* (Madrid, 1599).
24. P. Herrera Puga, 'Enfermedad y prostitución en la Sevilla de los Austrias', *Actas, IV Congreso Español de Historia de la Medicina*, **1** (Granada, 1975), 125–33.
25. *Fuentes legales de la medicina española (siglos XIII–XIX)*, ed. R. Muñoz Garrido and C. Muñiz Fernández, (Salamanca, 1969), pp. 29–31. This collection of documents conveniently brings together petitions from the *Cortes* relating to medical affairs.
26. *Ibid.*, pp. 34–5. At the same session Philip criticised the universities of Salamanca and Valladolid for slackness in the examination of medical students and attempted to make this more rigorous by instructing them to require students to defend a thesis in public. Philip also influenced the medical studies of Spaniards through his ban of November 1559 on study at all universities outside the Peninsula except those of Naples, Rome and the College of Spain in the university of Bologna. Intended to prevent contact with European Protestants, the measure led to a sharp fall in the numbers of Spanish students, mostly from the Crowns of Aragón,

who went to study at Montpellier, a leading medical centre. Before Philip's ban 248 Spanish students matriculated at Montpellier in 1510–59; after the ban the figure for 1560–99 is just 12; D. Goodman, 'Philip II's Patronage of Science and Engineering', *British Journal for the History of Science*, **16** (1983), 50–2.

27. *Fuentes legales*, pp.42–3; *Cortes*, 1579, Madrid.
28. *Ibid.*, p. 29.
29. J. M. López Piñero, 'La disección y el saber anatomico en la España de la primera mitad del siglo XVI', *Cuadernos de Historia de la Medicina Española*, **13** (1974), 51–110. The university of Valencia, unlike the principal universities of Castile, was governed by the municipality, which also appointed the professors. A useful account of the university's unusual elevation of medicine above theology and law is given in López Piñero, *La Facultad de Medicina de la Universidad de Valencia. Aproximación a su historia* (Valencia, 1980).
30. T. Santander Rodríguez, 'La creación de la catedra de cirugia en la universidad de Salamanca', *Cuadernos de historia de la Medicina Española*, **4** (1965), 191–213. An assessment of Alcázar's contributions to surgery is given in the article on him in *Diccionario histórico de la ciencia moderna en España*, ed. López Piñero *et al.*, vol. 1 (Barcelona, 1983), pp. 34–6.
31. *Fuentes legales*, p. 46.
32. Philip's letters, sent simultaneously to the three universities, are reproduced in E. Esperabé Arteaga, *Historia pragmatica e interna de la universidad de Salamanca*, vol. 1 (Salamanca, 1914), pp. 612–3 and 615.
33. *Fuentes legales*, pp. 52–5.
34. *Ibid.*, p. 49.
35. Muñoz Garrido, 'Empiricos sanitarios españoles de los siglos XVI y XVII', *Cuadernos de Historia de la Medicina Española*, **6** (1976), 116.
36. AGS:PR 71/158 and 159 document the affair: Philip's decree of 26 January 1559, Valladolid; informative statements by Isabel Pérez de Peramato, the empiric's widow; a copy of the petition by *procuradores* of the *Cortes*, 20 September 1560, Toledo; and a copy of the recommendation of the remedy by the administrator and deputies of the Hospital Real de Corte, (n.d.).
37. AGS:GA 160/137. 'Relación de las medicinas que se embian a la ysla Tercera para el ospital de la ynfantería spañola que en ella ay de guarnición', (1584).
38. AGS:GA 341/281, 'Memoria de las medicinas que llevan las naos que salen fuera este año de 1591'. The oil was still in use in the nineteenth century; it was then stocked in the royal pharmacy at Aranjuez; J. L. Valverde López *et al.*, *La Botica del Real Sitio de Aranjuez* (Granada, 1979), p. 40 lists it as an item in an inventory of 1804.
39. IVDJ 91/473, Andrés de Noronha, 'Advertimi[ento] de cierto neg[ocio] sabido por orden de la Inquisición de Llerena'.
40. The forgery of correspondence of this type is persuasively argued in I. Loeb, 'La correspondance des Juifs d'Espagne avec ceux de Constantinople', *Revue des Etudes Juives*, **15** (1887), 262–76.
41. Sebastian's Act of 20 September 1568 ordered the university of Coimbra to receive 30 medical students of pure Old Christian lineage whose training was to be financed by taxes raised in the provinces. Philip III later reaffirmed the Act and extended it to finance the training there of trustworthy apothecaries; T. Braga, *Historia da Universidade de Coimbra nas suas relações com a instrucção publica portugueza*, vol. 2 (Lisbon, 1895), pp. 779, 783 and 810; J. Lúcio de Azevedo, *Historia dos Cristãos Novos Portugueses* (2nd ed., Lisbon, 1975), p. 167. Fear of Jewish physicians was not peculiar to the Peninsula; in 1610 the medical faculty at the university of Vienna alleged that Jewish physicians set out to kill one in ten of their Christian patients by prescribing the wrong drugs; S. Baron, *A Social and Religious History of the Jews*, vol. 11 (NY, 1967), p. 159.
42. J. Valverde and A. Llopis González, *Estudio sobre los fueros y privilegios del antiguo Colegio de Apotecarios de Valencia* (Granada, 1974), p. 35; the privileges of the College are listed on pp. 83–4.

43. Archivo de la Corona de Aragón, Barcelona, MS: 'Del actes y negocis del Collegí del Magnifichs Droguers de la Ciutat de Barcelona', *libre* 1, ff. 1–4v, viceroy's approval of the College's statutes and privileges, 5 November 1562, Barcelona.
44. M. Fernández-Carrión and J. L. Valverde, *Farmacia y Sociedad en Sevilla en el siglo XVI* (Seville, 1985), pp. 14–15.
45. Juan Benito Guardiola, *Tratado de nobleza, y dictados que oy día tienen los varones claros y grandes de España* (Madrid, 1595), pp. 9v–10.
46. Granjel, *La medicina española renacentista, op. cit.*, p. 80.
47. J. Simon Díaz, 'La Inquisición de Logroño (1570–1580)' *Berceo. Boletín del Instituto de Estudios Riojanos* (Logroño), **1** (1946), 94.
48. L. García Ballester, *Historia social de la medicina en España de los siglos XIII al XVI. La minoría musulmana y morisca* (Madrid, 1976), p. 117.
49. A. Domínguez Ortiz and B. Vincent, *Historia de los Moriscos* (Madrid, 1978), p. 121.
50. *Fuentes legales*, pp. 195–6; Pedro de Vesga at the *Cortes*, Madrid, 1607 urging Philip III to prevent Moriscos graduating in medicine or even attending any medical lectures, supported his case with two recent confessions extracted by the Inquisition: a Morisco physician known as 'the avenger' who confessed to killing 3048 persons by poison; and a Morisco bone-setter who confessed to maliciously maiming Old Christians so that they would be unable to bear arms.
51. *Ibid.*, pp. 38–9.
52. M. Parrilla Hermida, 'Un título de cirujano en 1579', *Asclepio*, **25** (1973), 173–8 reproduces the text of the licence.
53. J. Valverde and R. García Serrano, *Colección documental de interés histórico-farmacéutico conservada en el Archivo General de Navarra* (Granada, 1979), pp. 22–3.
54. *Ibid.*, p. 24.
55. *Ibid.*, pp. 28–9.
56. For the inspection, supposed to be made every two years, of apothecaries' shops within five leagues of Madrid the *protomédico*'s inspector received 748 *maravedís* a day, an accompanying apothecary 500, and a notary and *fiscal* 300 each; *Recopilación de las leyes destos reynos* (Alcalá, 1598), *lib.* iii, *tít.* xvi, *ley* vii, Philip II, 1588, Madrid.
57. Archivo Real y General de Navarra, Pamplona: *sección de 'Medicina Cirujía, etc.', legajo* 1, *carpeta* 7 contains the physicians' petition to the *Cortes* of Navarre, and the decisions of the *Cortes* on 11–12 May 1586.
58. F. Oliver Rubio and F. Zubiri Vidal, 'Un códice del siglo XVI', *Archivos del Estudios Médicos Aragoneses*, **4–5** (1957), 271–94.
59. J. Galindo Antón, 'El arte de curar en la legislación foral aragonesa', *Actas III Jornadas Médicas Aragonesas* (Zaragoza, 1959), p. 356.
60. R. Jordi González, *Relaciones de los boticarios catalanes con las instituciones centrales* (Barcelona, 1975).
61. *Ibid.*, pp. 60 and 67.
62. *Ibid.*, p. 67.
63. *Ibid.*, p. 62.
64. This was one of the earliest pharmacopoeias; the only one to have appeared before then was that of Florence in 1498. The royal approval given to the 1587 edition was no more than a gesture; the *protomédico* had no part in its compilation.
65. Jordi González, *op. cit.*, p. 281 f.
66. *Ibid.*, p. 69.
67. J. López Piñero, 'Valencia y la medicina del Renacimiento y del Barroco', *Actas del III Congreso Nacional de Historia de la Medicina*, **2** (Valencia, 1971), 101.
68. Philip's letters of appointment of Collado and Coçar were published in *Solemne sesión apologética celebrada por la facultad de medicina de Valencia para honrar la memoria de sus antiguos catedráticos los doctores Plaza, Collado y Piquer* (Valencia, 1895), pp. 107–9.
69. Valverde and Llopis, *Colegio de Apotecarios de Valencia*, p. 35 f.
70. *Fuentes legales*, pp. 47–9, pragmatic of Philip II, 1588, Madrid. The same decree doubled the fine for unauthorised practice to 6000 *maravedís*.

71. *Ibid.*, pp. 50–1, pragmatic of Philip II, 2 August 1593, Escorial.
72. The king to Luis Mercado, 20 September 1593, Escorial, printed in the foreword to Mercado's *Institutiones chirurgicae iussu regio factae pro chirurgii in praxi examinandis* (Madrid, 1594).
73. Mercado, *Institutiones medicae iussu regio factae pro medicis in praxi examinandis* (Madrid, 1594). The use of this was discontinued by Philip III because learning it by heart caused 'such labour that the rest of medicine was neglected', particularly the Hippocratic and Galenic doctrines on 'fevers, pulses, purges, prognosis, aphorisms, places and effects', not included in Mercado's text; *Fuentes legales*, p. 57; pragmatic of Philip III, 17 November 1617, el Pardo.
74. *Instituciones que su magestad mando hazer al Doctor Mercado, su Médico de Cámara y Protomédico general, para el aprovechamiento y examen de los algebristas* (Madrid, 1599).
75. Francisco de Valles, *Tratado de las aguas destiladas, pesos, y medidas de que los boticarios deven usar, por nueva ordenanza y mandato de su Magestad y su Real Consejo* (Madrid, 1592), p. 43 v.
76. *Fuentes legales*, pp. 188–9; the protest came from the apothecaries of Madrid. As for their training, Philip's decree of 1563 required the apothecaries of Castile to serve four years' apprenticeship with a licensed practitioner, and to acquire a knowledge of Latin, so that they could interpret physicians' prescriptions; *ibid.*, p.35.
77. My comments on the numbers and salaries of royal physicians are based on the data tabulated in J. M. Jiménez Muñoz, 'Salarios de médicos y cirujanos: Nóminas de Corte (1538–1600)', *Asclepio*, **33** (1981), 315–34.
78. AGS:*Casa Real (Sitios Reales), legajo* 280, *apartadijo* 4, folio 1139, petition by *licenciado* Fragoso, 9 February 1580.
79. AGS:GA 137/107, *consulta*, *Consejo de las Indias*, 10 March 1582, Madrid. The sums in question here, especially that authorised for Valles, must have been for special services rendered, and perhaps in part were a gratuity; they are far too large to have come from the arrears of pay of a *médico de cámara* and *protomédico*.
80. AGS:E 13/73, 'Los médicos, cirujanos en quien paresce que concurren las calidades que se requieren pa[ra] servirse dellos de su magd', undated.
81. ARP:*Administrativa, Dependencias de la Real Casa*: *'Botica y Laboratorio'*, *legajo* 429, evidence taken in December 1590, Madrid. The witnesses included Álvarez de Pérez and *licenciado* Soto, physicians to the king's servants. Another document within this *legajo*, containing draft regulations by Mercado and two other royal physicians for the new *botica real*, is reproduced in J. de la Vega y Portilla, 'La botica real durante la dinastica austriaca', *Anales de la Real Academia de Farmacia*, **12** (1946), 380–4.
82. Some are illustrated in J. Hossard, 'La pharmacie de l'Escorial: ce qu'elle fut, ce qui reste', *Revue d'histoire de la pharmacie*, **15** (1961–2), plate 15; others in P. Herrero Hinojo and S. Muñoz Calvo, 'Boticas y enfermerías en los monasterios Jerónimos', *Studia Hieronymiana*, **2** (1973), 465–84.
83. AGS:DC 46/23, documents of draft contracts and advisors' comments, 1557-8. Mendoza asked for the negotiations to be held 'in great secrecy' to prevent the Portuguese learning of this threat to their spice trade. In the 1570s attempts were made in Seville to acclimatise ginger roots sent from the Indies; IVDJ 72/118, marqués del Carpio to the king, 1 August 1574. Substantial quantities of the spice were imported to Spain, mostly from Hispaniola, over 6000 *quintales* in 1584; E. Lorenzo Sanz, *Comercio de España con América en la época de Felipe II*, vol. 1 (2nd edition, Valladolid, 1986), pp. 606–8.
84. AGS:*Casa Real (Sitios Reales), legajo* 252, part 3, f. 125, Alonso de Mesa to Pedro de Hoyo, 16 July 1566, Aranjuez. The preparations listed here as ready for the king's physicians to prescribe included essences of fennel, chicory, marjoram and roses. The same document reveals that royal physicians sometimes visited Aranjuez and gave instructions to Holbeeck: Dr Mena asked him to prepare extracts from myrtle. Seville was one of the places Holbeeck travelled to in order to buy

herbs to plant in the gardens of Aranjuez, ARP:*Cédulas Reales*, vol. 3, ff. 4v–5r, the king to Francisco de Castilla, *asistente* of Seville, 20 December 1565, Madrid.

85. ARP:*Cédulas Reales*, vol. 2, f. 450v, the king to justices and other local officials in Castile, 17 May 1565, Escorial.
86. *Ibid., ibid.*, vol. 3, f. 256v, the king to the paymaster of the *alcázar* of Madrid, 11 January 1580, Madrid; IVDJ 99/302, Cardinal Granvelle to Mateo Vázquez, 21 October 1581, Madrid. Forte's salary of 60 000 *maravedís*, equal to that of a royal physician, was soon raised by 2 *reales* a working day. Holbeeck was paid more than this, though his duties were greater, involving care of the gardens as well as distilling: his initial salary of 112 500 *maravedís* was later increased by grants of 60 *fanegas* of wheat, 40 of barley, and an additional 3 *reales* a day to pay for house servants; ARP: *Cédulas Reales*, vol. 8, ff. 443–4, the king to the governor of Aranjuez, 6 August 1594, Escorial.
87. Bartolomé de Las Casas, *Apologética historia sumaria*, ed. E. O'Gorman, vol. 1 (Mexico, 1967), pp. 97 and 106; the work is thought to have been written in 1555–9. The use of guaiacum had already spread to central Europe in the 1510s. Its effectiveness against syphilis, entirely imaginary, was asserted for two centuries; the supposed benefits were probably due to the avoidance of complications caused by the alternative remedy, mercury, which is toxic. The quantities of guaiacum, liquidambar, sarsaparilla, china-root, cassia fistula and mechoacan – all medicinal plants – shipped to Spain from the Indies during various years of the reign are tabulated in Sanz, *op. cit.*, vol. 1, pp. 604–13.
88. L. Benítez Miura, 'El Dr Francisco Hernández: 1514–1578 (cartas inéditas)', *Anuario de Estudios Américanos*, **7** (1950), 398; Hernández to the king, 30 April 1572, Mexico City.
89. *Cedulario indiano, op. cit.*, vol. 1 (Madrid, 1596), p. 224; the king to Hernández, 11 January 1570, Madrid. Hernández's letters from New Spain to the king in 1572–6 are published in Benítez Miura, *op.cit.*, pp. 396–408; J. Toribio Medina, *Biblioteca Hispanoamericana (1493–1810)*, vol. 2 (Chile, 1900), pp. 272–90; and Codoin, 1, pp. 362–79. A detailed account of Hernández's travels in New Spain and the complicated publication of his works is given in G. Somolinos D'Ardois, *Vida y obra de Francisco Hernández* (Mexico, 1960).
90. An inflated title since the king made it clear that his powers to license practitioners were restricted to five leagues from his place of residence; nor was he free to remove licences already given by other authorities. Philip had appointed a *protomédico* of Peru in 1568: Sánchez de Renedo, whose jurisdiction similarly observed the Castilian precedent of a five-league zone. As for New Spain, control over medical practitioners in Mexico City was in the hands of the municipal authorities. Hernández's attempts to inspect pharmacies in the city led to quarrels, and he was prevented from parading in the streets preceded by a servant bearing his staff of office. Trouble again arose, after Hernández had left, when in 1585 the viceroy tried to impose a royal *protomédico* on the city. It was an American manifestation of a conflict which had occurred frequently in the Peninsula; the difference was that by 1603 Mexico City had given up the fight to preserve powers which were established on a much weaker base. The conflict is described in J. Lanning, *The Royal Protomedicato: the Regulation of the Medical Professions in the Spanish Empire* (Durham, NC, 1978), pp. 27–8; this study is largely concerned with the eighteenth century.
91. IVDJ 99/190–2, Juan de Herrera to Mateo Vázquez, 5 May 1582.
92. A tabulated listing of the species described, their therapeutic action, and their supposed degrees of hot, cold, moist and dry – the four classical qualities by which Hernández continued to explain their medicinal activity – is presented in M. Sánchez Tellez, F. Guerra, and J. Valverde, *La doctrina farmacéutica del Renacimiento en la obra de Francisco Hernández, c. 1515–1587* (Granada, 1979).
93. *Relaciones geográficas de Indias: Perú*, ed. M. Jiménez de la Espada, vol. 2 (new edition, Madrid, 1965), p. 239; *relación* of Otavalo prepared by Sancho de Paz

Ponce de León, *corregidor*, 2 April 1582. The *corregidor* added that information on all of the herbs could be supplied by Dr Heras, a herbalist who had 'recorded descriptions of most of them'.

94. *Gobernantes del Perú*, ed. R. Levillier, vol. 8 (Madrid, 1924), p. 28; ordinances of Viceroy Toledo, 3 October 1572, Cuzco.
95. F. del Paso y Troncoso, *Papeles de Nueva España: relaciones geográficas de México* (reprinted, Mexico City, 1979), pp. 61–2; *relación* of Coatepec, 1579. This report reveals that Spaniards were travelling to this town from Mexico City to consult Indians about the virtues of the local plants.
96. Bibliothèque Nationale, Paris; *Manuscrits Portugais* no. 2, ff. 29–79, 'Esperiencias das hervas orientales que sua Magd. mandou fazer ao vizorey Mathias de Albuquerque en anno de 1596'.
97. IVDJ 21a/31, conde de Priego to unnamed royal secretary, October 1596, Seville. The *conde*, apparently in reply to the king's specific inquiries, stated that while much of the garden's stock could be transplanted, the trees 'of which there are several rare specimens' would not survive the move. Also that the garden, which measured 192 feet by 295, would cost no more to maintain than one gardener's wage. He advised the king to consult the royal cosmographer Zamorano – he also had formed a plant collection – for advice on the garden's management and on which other plants should be introduced there. The creator of the garden, Tovar, had written works on the composition of medicines and assisted in the inspection of pharmacies in Seville.
98. ARP:*Administrativa: Bosques y Montes, legajo* 335, Juan de Ibarra, secretary of the *Junta de Obras y Bosques* to the king, 26 March 1598, Madrid; ibid., *Cédulas Reales*, vol. 9, ff. 318v–319r, the king to the paymaster of works of the *alcázar* of Madrid, 24 April 1598, designating Pomar's appointment with an annual salary of 60 000 *maravedís*. The purchase of Diego de Burgos' house and garden for 2200 ducats in January 1566 allowed the royal apothecary to live in the house and continue cultivating the garden until his death; ARP: *Administrativa*: *Títulos*, *legajo* 1212, *expediente* 33, the king to Pedro de Santoyo, paymaster of works of the alcázar of Madrid, 15 January 1566, Madrid. This garden, along with other royal parks and residences, and the mint of Segovia, was administered by the *Junta de Obras y Bosques* (Committee of Works and Woods), a body established in 1545; its functions and secretaries up to 1621 are recorded in IVDJ 100/344.
99. AGS:GA 96/111, the king to marqués de Santa Cruz, draft letter, 31 January 1580, Madrid. This was soon reported to have been done. Philip's orders also instructed Santa Cruz to send a list of their names to justices of the towns concerned, asking them to make sure that none of the sick left until they had fully recovered in the hospitals where they were being treated, by which time it would be safe for them to rejoin the armada.
100. *Ibid.*, 96/53, same to the same, January 1580, Madrid. As if this were not enough, there was pestilence in Gibraltar with reports of 300 deaths and 2000 others sick; Santa Cruz therefore ordered the galleys of Naples, bringing German and Spanish troops, not to call there but instead to proceed directly to Cádiz; he wrote: 'I hope to God the men will be in good health to serve Your Majesty whenever they are needed, and that the sick return to their stations as they recover', *ibid.*, 102/12, Santa Cruz to the king, 3 March 1580, Puerto de Santa María.
101. *Codoin*, 33, p. 265, the king to the duque de Alva, 28 November 1580, Badajoz. The precautions were insufficient; a fortnight later the duke of Alva, commander of the army, reported that the infection had entered the castle. The duke summoned a meeting of physicians and municipal officials; the recommendations, approved by Philip, included the selection of a district to isolate the sick and financial assistance from the crown. Philip's instructions to clean up the city were soon implemented. Alva used galley convicts for the task, though he believed that 'the principal and most certain remedy' would come from Philip's other order for prelates to pray to God to placate His anger; *ibid.*, pp. 257; 272; 284; corres-

pondence between Alva and the king, November – December 1580. Alva advised Philip against coming to Lisbon to receive the crown of Portugal – the ceremony was later performed at Tomar instead. But Philip caught the infection and, as it spread to Castile, his wife Ana died from it.

102. *Ibid.*, p. 464, the king to the duque de Alva, 21 January 1581, Elvas.
103. AGS:GA 159/125, Francisco Benito de Mena to Jorge Manrique, *veedor general* of the armada, 12 January 1584, Jerez. The reports of pestilence were not entirely groundless. After the sceptical de Mena sought to discover the truth by consulting the local physicians it was learned that two had died from the contagion and one other was sick; *ibid.*, 159/113, 'lo que dicen el doctor Aleman y el *licenciado* Clavo, médicos de la ciudad de Xerez sobre la enfermedad que an publicado ay en ella', 18 January 1584, Jerez. Although there were no more cases than this, one wonders how reassuring this was to Manrique, the *veedor* (inspector) of the armada, who had expressed anxiety for the health of infantry stationed on galleys just two leagues from Jerez.
104. *Ibid.*, 151/126–7, city of Cádiz to Juan Delgado, secretary of the *Consejo de Guerra*, 17 November 1583, Cádiz; with testimony of Martín de Yrigoen, *regidor* and *procurador mayor* of Cádiz; Diego de Cuellar and Diego Arias, physicians of Cádiz; and four surgeons of the city, 16 November 1583, Cádiz.
105. *Ibid.*, 283/306, Juan Blan Rivera to the king, 12 April 1590, Perpiñán.
106. *Ibid.*, 84/85, *consulta*, *Consejo de Guerra*, 13 August 1578, Madrid, responding to the request of Tomás López, administrator of the royal hospital of Villafranca de Montesdoca.
107. *Ibid.*, 140/139–40, majordomo and officials of the Hospital de la Misericordia to Antonio de Eraso, royal secretary, 20 March 1582, San Lúcar, with evidence taken in July 1576 of the large influx of sick soldiers and seamen.
108. *Ibid.*, 227/25–6, archbishop of Santiago to the king, 23 September 1588, Santiago de Compostela, reporting on medical assistance given from 3 July.
109. *Ibid.*, 236/351, the king to Francisco de Arriola, *contador* of the galleys of Spain, 17 November 1588, Madrid, ordering an assessment of compensation to be paid to the owners of the destroyed houses; with Arriola's reply, 12 December 1588, San Sebastián; and *ibid.*, 264/214, Arriola, 'Relación de lo que montan las medicinas que se an gastado en la ospitalidad de los enfermos que vinieron en las naos de la armada que aportaron a los veinte e tres de setiembre del año passado de quinientos y ochenta y ocho al puerto del Pasaje', 28 September 1589, San Sebastián.
110. S. Montserrat, *La medicina militar á través de los siglos* (Madrid, 1946), p. 182.
111. AGS:GA 277/230, Dr Espinosa to the king, 2 January 1589, Salamanca.
112. MN:FN III/6/136v, the king to Juan de Mendoza, captain-general of the galleys of Spain, 19 August 1557, St Quentin. The cost of providing medicines (and hens) for a single galley was estimated at 75 000 *maravedís* a year, just under four per cent of the total cost of maintenance: *ibid.*, XII/84/315v, 'Relación del gasto que una galera hace en un año', undated.
113. AGS:GA 269/94, petition of Dr Quesada to the *Consejo de Guerra*, 9 March 1589; *ibid.*, 301/199, petition of Dr Guevara for an increase in pay, 1590.
114. *Ibid.*, 78/120, *consulta*, *Consejo de Guerra*, 1574.
115. *Ibid.*, 209/120, *consulta*, *Consejo de Guerra*, 17 August 1587, Madrid, recommending a donation of 200 ducats in response to the appeal by Diego de Vivero, *contador* and majordomo of the military hospital of Pamplona.
116. *Ibid.*, 172/156, the king to Luis de Arteaga, *corregidor* of the cities of Murcia, Lorca and Cartagena, 1 June 1584, Escorial; and Luis de Arteaga and Juan de Escobedo Riba de Neyra to the king, 3 July 1584, Cartagena.
117. *Ibid.*, 185/135, the *adelantado* of Castile, captain-general of the galleys of Spain, to the king, 29 April 1586, Cádiz; and *ibid.*, 188/57, same to the same, 9 October 1586, Gibraltar.
118. L. van Meerbeeck, 'Le service sanitaire de l'Armée espagnole des Pays-Bas à la

fin du XVIme et au XVIIme siècles', *Revue internationale d'histoire militaire,* **5** (1956), 479–93; and G. Parker, *The Army of Flanders and the Spanish Road 1567–1659* (Cambridge, 1972), pp. 167–8.

119. AGS:GA 218/52, Pedro de Padilla, administrator of the hospital of San Bernardino, to the king, 18 September 1587, Oran; *ibid.*, 209/35, *consulta, Consejo de Guerra,* 16 October 1587, Madrid; *ibid.*, 213/504, petition for pay by Pedro Crietes, physician to the soldiers of Oran and Mers-el-Kebir, c. 1586: *ibid.*, 284/342, duque de Cardona, governor of the fortresses, to the king, 26 May 1590, Oran, reporting the continuing financial difficulties of the hospital.
120. *Ibid.*, 242/217, petition by the majordomo of the hospital at Peñón de Vélez, sent to the Council of War, c. 1587; *ibid.*, 277/275, petition to the crown by Cristóbal de Aguilar de Viana, no date.
121. *Ibid.*, 202/160, Francisco Duarte to the king, 8 October 1587, Lisbon; and 203/18, Jorge Manrique to the king, 15 November 1587, Lisbon.
122. *Ibid.*, 251/137, Francisco Coloma to the king, 16 September 1589, Lisbon.
123. Several have recently been identified by M. Gracia Rivas, 'El personel sanitario que participó en la jornada de Inglaterra. Nuevas aportaciones', *Revista de Historia Naval,* **1** (1983), 63–90.
124. Francisco de Reynoso served on the expedition as physician and surgeon-major to the *tercio* commanded by Antonio Pereira, and when the Armada returned he assisted at the improvised hospital in Santander; AGS:GA 312/240, *licenciado* Francisco de Reynoso to the king, 19 February 1590. Pedro Suárez went as apothecary to the artillery; the medicines he took for that purpose were subsequently made available to the Armada hospital to make good the loss, which occurred during the expedition, of medicines to the value of 2000 *escudos*; *ibid.*, 274/244–5, statement by Suárez with official appraisal of his services and comments by Dr Valles, 3–4 July 1589. Two other practitioners attached to specific units are mentioned in note 125.
125. *Ibid.*, 236/290, medical certificate declaring Sebastián de Prieda, soldier of the company of Captain Pantoja, unfit for duty; issued by Drs Giuseppe Goncalvez de Counedo, physician to the *tercio* of Sicily and Lucas Cerdan, physician to the artillery of the Armada, 8 July 1588, La Coruña; with Medina Sidonia's authorisation for discharge.
126. *Ibid.*, 323/169, Manuel Lobo de Andrada to the king, 8 July 1591, Blavet; and 371/150b, Juan de Pedroso to the king, 3 January 1593, Blavet. In its first four months of functioning the records show that there were 557 admissions, mostly suffering from illnesses 'due to the cold', some of them dangerously ill. They were treated with ointments and put in hothouses to induce sweating. Of this number 56 died, 45 were still in hospital and 456 had been discharged after treatment; *ibid.*, 341/287, 'Relación de la gente de guerra y mar que se ha curado en el hospital real de Blavet', 1591.
127. Their names and salaries are listed in *ibid.*, 341/38, 'Relación de todo lo que se a gastado en este real ospital de la armada', December 1590, El Ferrol.
128. Letters from Dr Manso to the king, July–October 1593, Huesca and Zaragoza, *ibid.*, 375/4 and 19; 377/90 and 378/44.
129. AGS:E 1124/260, Antonio Monteflor (called 'Padre Grego') to the king, 3 October 1559, Messina, with accompanying diagram of his hospital; *ibid.*, 1124/225, duque de Medinaceli to the king, 19 August 1559, Messina, refers to the high opinion of the design.
130. AGS:GA 73/124–6, 'Relación de los soldados enfermos que se hallan en este hospital del exercito de su mgt. en los Padules', 19 May 1570.
131. *Codoin,* 32, p. 93, duque de Alva to Juan Delgado, secretary, *Consejo de Guerra,* 23 April 1580, Llerena; M. Parrilla Hermida, 'La Anexión de Portugal en 1580: el hospital de campaña', *Asclepio,* **28** (1976), 275–8.
132. AGS:GA 149/21, *licenciado* Francisco de Ancona, physician, 'Relación de las dietas

que son necessarias en un año para provisión y servicio de los soldados heridos enfermos del ospital real desta ciudad de Angla [sic]', 1 October 1583, Angra.

133. *Ibid.*, 282/346, Luis de Benavides to the king, 21 March 1590, Santa Cruz de Tenerife. The king provided 200 ducats for this physician's salary, *ibid.*, 287/297, Luis de la Cueva to the king, 30 August 1590, Canaria.
134. *Ibid.*, 237/235, Dr Valles to Andrés de Alva, secretary of war (sea), 3 February 1588, royal palace, Madrid; *ibid.*, 268/147–9, Dr Valles to Andrés de Prada, secretary of war (land), 6 and 14 June 1589, Escorial.
135. *Ibid.*, 174/138, draft royal *cédula* appointing Pérez de Herrera to replace Juan de la Fuente, the deceased *protomédico* of the galleys of Spain, 1584.
136. F. Pascarella, 'Un documento inédito del siglo XVI sobre el nombramiento del cirujano mayor del ejercito español en la Italia septentrional', *Archivo Iberoamericano de Historia de la Medicina*, **8** (1956), 315–20.
137. AGS:GA 74/90, petition by Rodrigo de Molina to the *Consejo de Guerra* for salary owing to him, undated.
138. *Ibid.*, 181/406, petition to the *Consejo de Guerra*, 1585.
139. *Ibid.*, 308/163, petition to the *Consejo de Guerra*, 1590. There was later a request from the Basque coast for the employment of a native physician to accompany an armada, because recent experience had shown that 'deaths had been caused' by the unavailability of naval physicians able to understand the Basque language; *ibid.*, 319/131, Antonio de Urquiola to the king, 17 March 1591, Pasajes.
140. This and his theory of disease are revealed in his defence against legal proceedings taken against him in Madrid by the king's attorney Martín Ramón for using unorthodox and unapproved medicines; BM Add. MS. 28,353, ff. 57–61, undated.
141. AGS:GA 341/286, 'Memoria de las drogas que son menester en la botica real de la armada de su magd. para componer las medicinas, necesarias para la cura de la gente de mar y guerra por espacio del año', 8 March 1591, El Ferrol. The list did not include medicines already in stock and therefore does not give a full indication of the pharmacy's requirements for the year.
142. *Ibid.*, 341/ 288, 'Relación de las medicinas simples y compuestas y dietas que se an embarcado en los felibotes del cargo del Capitán Pedro de Zubiaurre de respeto para llevar a Bretaña para servicio y regalo de los enfermos que ay en el ospital que su magd. tiene en aquel reyno', 6 September 1591, El Ferrol; *ibid.*, 341/63, 'Relación de lo que se a entregado a Antonio de Viana mayordomo del ospital que va al rrey[no] de Bretaña', 25 February 1591, El Ferrol. The second document indicates that pharmaceutical instruments were also sent, including a special large strainer for making preparations from cassia fistula; 72 glazed earthenware vessels for the pharmacy were put on board. Cassia fistula (12 *libras*), guaiacum (12 *libras*) and sarsaparilla (20 *libras*) were also supplied to the hospital of the infantry garrisoning Terceira; *ibid.*, 160/137, 'Relación de las medicinas que se embian a la ysla Tercera para el ospital de la ynfantería española que en ella ay de guarnición', 1584.
143. But elder oil, previously used as a boiling-hot application for wounds, and linseed oil, another poultice, are also listed. Does this mean that the older drastic treatment of wounds had not completely disappeared; or simply that these oils were used for other purposes?
144. AGS:GA 341/38, 'Relación de todo lo que se a gastado en este Real ospital del armada', December 1590, El Ferrol.
145. *Ibid.*, 291/115, annotation by Andrés de Alva, secretary of the *Consejo de Guerra*, to Dr Manso's letter; he commented that the king 'is very well served' by the charitable work of the hospital and the gain to the treasury (meaning probably the recovery of paid soldiers and seamen rather than the economy of hospital expenditure).
146. *Ibid.*, 111/105, Pedro Bermudez to the king, 16 March 1581, Lisbon. A similar suggestion was later made by the king for sick seamen at La Coruña, *ibid.*,

381/220, the king to Martín de Ayala, 22 May 1583, Aranjuez. The formation of a second armada hospital at Lisbon was intentionally avoided in 1592 for reasons of expense; instead arrangements were made for the sick to be treated by religious brethren of the Order of San Juan de Dios then caring for soldiers at the castle of Lisbon; *ibid.*, 353/93, Bernabé de Pedroso to the king, 20 June 1592, Lisbon.

147. *Ibid.*, 186/93, Cristóbal de Heredia, *proveedor* of Cartagena, to the king, 26 July 1586; and 213/559, petition of Juan de la Rubia to the *Consejo de Guerra*, 1586.

148. *Ibid.*, 375/4, Dr Manso to the king, 1 July 1593, Huesca.

# Conclusion

Scientific and technological activity in Philip II's Spain were strongly influenced by government policy. The centralising tendencies of the crown, the control of empire and the needs of war were all important in this. The hand of government was felt in the university training of physicians and surgeons, in the technical training of gunners and pilots, and in the collection of scientific information concerning the Indies. To achieve its ends the crown was forced to rely on the knowledge and skills of men from parts of Europe where technologies were more developed than in Spain. But it was the crown's intention that this reliance on foreigners should be temporary, and on several occasions royal officials were instructed to make arrangements to promote this policy by encouraging the communication of technical skills from the foreigners to the natives. At Barcelona, the centre of galley construction, the crown expected Spanish craftsmen to learn from the Genoese; and at Guadalcanal to acquire advanced mining techniques from the Germans. And Philip's Academy of Mathematics in Madrid had been created primarily to secure a native supply of engineers. The plans failed and in the seventeenth century Spain continued to depend on foreign technicians. Nevertheless there are clear signs during Philip II's reign of important Iberian contributions to mining (Enrique Garcés, Pedro de Contreras, Pedro Fernández de Velasco); of these, Garcés' work at least seems to have been stimulated by informal contacts with Germans in the Peninsula, just as Bartolomé de Medina's all-important adaptation of the amalgamation process.

Why were the government's plans no more successful? The policy of intolerance cannot be ignored here; the crown continued to deprive itself of the full resource of native talent by the severe disabilities imposed on the New Christian population which excluded them from university education and numerous occupations. But this has to be qualified by the recognition that the bans were sometimes relaxed. On several occasions when the crown needed the services of these men their origins were overlooked and they were employed. *Converso* physicians and surgeons served the crown. And at Guadalcanal a striking example of crown compromise has been noted in the employment of a Morisco foreman, Francisco Blanco; although sentenced to exile by the Inquisition, his mining skills mattered more and aroused great hopes of boosting royal silver production. Was there a failure of response from the Old Christian population to government appeals for native technicians? Did Castilians reject the call because they disparaged surgery, mining, gunnery as unworthy, menial occupations? The traditional image of the Castilian as averse to commerce is being eroded by recent research which shows their considerable participation in the trade of the Indies. The associated attribution of Castilian aversion to technology may well have to be reconsidered as well. Viceroy Toledo's observation that in Peru the Spaniards preferred to become soldiers rather than proprietors of mines has to be taken seriously, but in the Indies and the Peninsula there is evidence of intense interest in mining whenever there were prospects of quick wealth; and in the Peninsula local labourers on occasion volunteered for work in the pits, as did the population of Azuaga in the first weeks of exploitation at Guadalcanal. But interest and initiative in mining were stifled by the crown's expropriation of mines and the resented royal duty. Poor economic rewards may well have been the main reason for the crown's shortages in military physicians, pilots and gunners; more important than any considerations of dishonour associated with these occupations. At any rate there is abundant evidence in the documents of royal officials appealing to the king to raise salaries to attract skilled men to these much-needed tasks. If the king was hesitant to grant this it was because he could not afford to do so. And shortage of money has been shown to be a constantly recurring theme in all matters relating to crown technological projects; at times the crown was at a loss to find even small sums of money owing to individuals whose services it had employed. The debate over whether manufacturing or provisioning should be directly administered by the crown or left to private individuals arose in mining, shipbuilding and the procurement of pharmaceutical supplies for the armed forces. The preference for royal administration in the interests of greater crown control often gave way to the other course because of inadequate royal finances. The failure of the treasury, drained by inherited debts and expanding warfare, was the most important reason for Spain's limited technological achievement. But to this must be added the undeveloped state of contemporary technology; nowhere in the world would it have been possible at the time to have found solutions to some of the problems tackled: the efficient ventilation and drainage of pits; the provision of potable water on board ships; the manufacture of strong and accurate cannon.

The king's personal interests in science and technology have also to be considered because of his intimate involvement in all the affairs of government. His dilettantism, reflected in his interests in astrology, Lullian philosophy, water-divining, alchemy, unconventional medicine, and the virtues of plants led to the patronage of numerous natural philosophers and inventors. His was anything but a closed mind. He was prepared to consider the new and unconventional, and this attitude was reflected in his advisory councils to whom the stream of inventors were generally referred. Any project which seemed to be in the interests of Spain's power or wellbeing was given a formal trial; an empirical approach sobered by experience of tricksters and impossible projects. Some innovations did not survive the death of the monarch: the alchemical and other scientific activity at the Escorial was the chief casualty here. But his Academy of Mathematics lasted another two decades. More durable were the king's mining regulations which remained in force into the nineteenth century.

The achievements of the reign which stand out most of all are the restoration of the fleet through the programme of shipbuilding; the collection of scientific information from the Indies, especially the Mexican flora of Francisco Hernández; the organisation of medical facilities for the military; the organisation and exploitation of American mines. It is difficult not to be impressed by the boldness and scale of some of these enterprises; and by the determination, perseverance and dedication of the royal officials who were given charge of them: Cristóbal de Barros in Peninsular shipbuilding; López de Velasco in the *relaciones* of the Indies; Francisco Hernández's indefatigable survey of the natural history of New Spain; Francisco de Mendoza's supervision of mining within the Peninsula and Viceroy Toledo's organisation of mining in Peru.

Their examples alone suffice to weaken the traditional image of Castilians as uninterested in technology and science. But recurring financial crises continued to impede Spain's technological development. And the obstruction of royal technological projects by various individuals and local authorities not only in the Indies but also in the Peninsula provide further evidence that Philip II's power was far from absolute.

# Bibliography of printed sources

## (a) Primary sources


Almela, J. de, *Descripción de la octava maravilla del mundo*, ed. Andrés, G. de, *Documentos para la Historia del Monasterio de San Lorenzo el Real de El Escorial*, **6** (1962), 5–98.

Bleiberg, G., (ed.), *El "Informe secreto" de Mateo Aleman sobre el trabajo forzoso en las minas de Almadén* (London, 1985).

Cabrera de Córdoba, L., *Historia de Felipe Segundo, rey de España* (Madrid, 1619).

Capoche, L., *Relación general del asiento y Villa Imperial de Potosí*, ed. Hanke, L., (Madrid, 1959).

Cárdenas, J. de, *Problemas y secretos maravillosos de las Indias* (Mexico, 1591).

Ciruelo, P., *A Treatise reproving All Superstitions and Forms of Witchcraft very necessary and useful for All Good Christians zealous for their Salvation*, trans. Maio, E. and Pearson, D'O., (London and NJ, 1977).

Coçar, L., *Dialogus veros medicinae fontes indicans* (Valencia, 1589; facsimile with introduction by López Piñero, J. M., Valencia, 1977).

*Colección de documentos inéditos para la historia de España*, ed. Fernández Navarrete, M., *et al.*, 112 vols., (Madrid, 1842–95).

*Colección de documentos inéditos relativos al descubrimiento, conquista y organización de las antiguas posesiones españoles de América y Oceanía*, ed. Pacheco, J. F., *et al.*, 42 vols., (Madrid, 1864–84).

*Colección de documentos para la historia de la formación social de Hispanoamérica 1493–1810*, ed. Konetzke, R., vol. 1 (Madrid, 1953).

*Colección documental de interés histórico-farmacéutico conservada en el Archivo General de Navarra*, ed Valverde, J. and García Serrano, R., (Granada, 1974).

Collado L., *Platica manual de artillería* (Milan, 1592).

*Correspondance du Cardinal de Granvelle*, ed. Poullet, E., and Piot, C., 12 vols., (Brussels, 1877–96).

Cortés, M., *Breve compendio de la sphera y de la arte de navegar* (Seville, 1551).

'Discurso á propósito del cometa de 1577 contra la astrología judiciaría', (Escorial MS. L-I-12, f.173), *La Ciudad de Dios*, **82** (1910), 292–8.

*Discurso de la vida del ilustrísimo y reverendísimo Señor Don Martín de Ayala escrito por sí mismo*, *Autobiografías y memorias*, ed. Serrano y Sanz, M., (Madrid, 1905).

'Discurso sobre el cometa que apareció en 13 de Julio de 1596', (Escorial MS. L-I-12, ff. 178–81), *La Ciudad de Dios*, **82** (1910), 190–4.

Encinas, D. de, *Cedulario indiano*, 4 vols., (Madrid, 1596).

Fernández del Castillo, J., *Tratado de ensayadores* (Madrid, 1623).

*Fuentes legales de la medicina española (siglos XIII–XIX)*, ed., Muñoz Garrido, R. and Muñiz Fernández, C., (Salamanca, 1969).

*Gobernantes del Perú*, ed., Levillier, R., 14 vols., (Madrid, 1921–6).
González, T., (ed.), *Noticia histórica documentada de los célebres minas de Guadalcanal desde su descubrimiento en el año de 1555, hasta que dejaron de labrarse por cuenta de la real hacienda*, 2 vols., (Madrid, 1831).
González, T., (ed.), *Registro y relación general de minas de la Corona de Castilla*, 2 vols. (Madrid, 1832).
Guadalajara y Xavierr, M. de, *Memorable expulsion y justíssimo destierro de los Moriscos de España* (Pamplona, 1613).
Guardiola, J. B., *Tratado de nobleza* (Madrid, 1595).
Horozco y Covarruvias, J. de, *Tratado de la verdadera y falsa prophecía* (Segovia, 1588).
*Instituciones que su magestad mandó hazer al Doctor Mercado, su Médico de Cámara y Protomédico general, para el aprovechamiento y examen de los algebristas* (Madrid, 1599).
*La Audiencia de Charcas. Correspondencia de presidentes y oidores*, ed., Levillier, R., 3 vols., (Madrid, 1918–22).
Las Casas, B. de, *Apologética historia sumaria*, ed. O'Gorman, E., (Mexico, 1967).
López de Velasco, J., *Geografía y descripción de las Indias*, ed. Jiménez de la Espada (Madrid, 1971).
Marmol Carvajal, L. del, *Historia del rebelión y castigo de los Moriscos del reino de Granada* (Madrid, 1858).
Medina, P. de, *Arte de navegar* (Valladolid, 1545).
Medina, P. de, *Regimiento de navegación* (Seville, 1563).
Mercado, L., *Institutiones chirurgicae iussu regio factae pro chirurgii in praxi examinandis* (Madrid, 1594).
Mercado, L., *Institutiones medicae iussu regio factae pro medicis in praxi examinandis* (Madrid, 1594).
Mercado, L., *Libro en que se trata con claridad la naturaleza, causas, providencia, y verdadera orden y modo de curar la enfermedad vulgar, y peste que en estos años se ha divulgado por toda España* (Madrid, 1599).
Morales, A. de., *Las antigüedades de las ciudades de España* (Alcalá, 1575).
Muñoz, J., *Libro del nuevo cometa* (Valencia, 1573; facsimile with introduction by Navarro Brotóns, V., Valencia, 1981).
*Novisima Recopilación de las leyes* (Madrid, 1614).
*Papeles de Nueva España*, ed. Paso y Troncoso, F. del, 7 vols., (Madrid and Paris, 1905–6).
Pérez de Moya, J., *Tratado de cosas de astronomía y cosmographía y philosophía natural* (Alcalá, 1573).
*Recopilación de las leyes destos reynos* (Alcalá, 1598).
*Recopilación de leyes de los reinos de las Indias* (Madrid, 1681).
*Relaciones geográficas de España y de Indias impresas y publicadas en el siglo XVI*, ed. Sanz, C., (Madrid, 1962).
*Relaciones geográficas de Indias: Perú*, ed. *Jiménez de la Espada*, M., 3 vols., (new edition, Madrid, 1965).
Rojas, C. de, *Teoría y práctica de fortificación* (Madrid, 1598).
Santa Cruz, A. de, *Libro de las longitudines y manera que hasta agora se ha tenido en el arte de navegar* (Seville, 1921).
Sigüenza, J. de., *Historia de la Orden de San Jerónimo* (Madrid, 1590).
Valles, F. de, *Tratado de las aguas destiladas, pesos, y medidas de que los boticarios deven usar, por nueva ordenanza y mandato de su Magestad y su Real Consejo* (Madrid, 1592).
Zamudio de Alfaro, A., *Orden para la cura y preservación de las secas y carbuncos* (Madrid, 1599).
Zarco, J., 'El hospital de El Escorial', *La Ciudad de Dios*, **132** (1923), 415–22; **133** (1923), 5–13 and 100–104. (Reproducing the MS. constitution of the hospital).

## (b) Secondary sources


Andrés, G. de, 'Entrega de la librería real de Felipe II (1576)', *Documentos para la Historia de San Lorenzo el Real de El Escorial*, **7** (1964), 7–233.

Andrés, G. de, 'Juan Bautista Gesio, cosmógrafo de Felipe II y portador de documentos geográficos desde Lisboa para la biblioteca de El Escorial en 1573,' *Publicaciones de la Real Sociedad Geográfica, serie* 'B', no. 478 (Madrid, 1967), pp. 1–12.

Antolín, G., 'La librería de Felipe II (datos para su reconstitución),' *La Ciudad de Dios*, **116** (1919), 36–49, 287–300, 477–88; **117** (1919), 207–17, 364–77; **118** (1919), 42–49, 123–37.

Arzáns de Orsúa y Vela, B., *Historia de la Villa Imperial de Potosí*, ed. Hanke, L. and Mendoza, G., (Providence, R. I., 1965).

Bakewell, P., *Silver Mining and Society in Colonial Mexico: Zacatecas (1546–1700)*, (Cambridge, 1971).

Bakewell, P., 'Registered Silver Production in the Potosí District 1550–1735', *Jahrbuch für Geschichte von Staat, Wirtschaft und Gesellschaft Lateinamerikas*, **12** (1975), 67–103.

Bakewell, P., 'Technological change in Potosí: The Silver Boom of the 1570s', *ibid.*, **14** (1977), 57–77.

Bakewell, P., 'Mining in Colonial Spanish America', *The Cambridge History of Latin America*, ed. Bethell, L., vol. 2 (Cambridge, 1984), pp. 105–51.

Bargalló, M., *La amalgamación de los minerales de plata en Hispanoamerica colonial* (Mexico, 1969).

Benítez Miura, L., 'El Dr Francisco Hernández: 1514–1578 (cartas inéditas)', *Anuario de Estudios Américanos*, **7** (1950), 396–408.

Bennassar, B., *Recherches sur les grandes épidémies dans le nord de l'Espagne à la fin du XVI^e^ siècle* (Paris, 1969).

Braudel, F., *The Mediterranean and the Mediterranean World in the Age of Philip II*, trans. Reynolds, S., 2 vols., (London, 1973).

Cabanelas Rodríguez, D., *El Morisco granadino Alonso del Castillo* (Granada, 1965).

Caro Baroja, J., *Vidas mágicas e Inquisición* (Madrid, 1967).

Caro Baroja, J., *Los Judíos en la España Moderna y Contemporánea*, 3 vols., (2nd edition, Madrid, 1978).

Carrera Pujal, J., *Historia de la economía española*, vol. 2 (Barcelona, 1944).

Carrera Pujal, J., *Historia politica y económica de Catalunya*, vol. 2 (Barcelona, 1947).

Destombes, M., 'Un astrolabe nautique de la Casa de Contratación (Seville, 1563),' *Revue d'histoire des sciences*, **22** (1969), 33–64.

*Diccionario histórico de la ciencia moderna en España*, ed. López Piñero, J. *et al.*, 2 vols., (Barcelona, 1983).

Domínguez Ortiz, A. and Vincent, B., *Historia de los Moriscos* (Madrid, 1978).

Duffy, C., *Siege Warfare* (London, 1979).

Esperabé Arteaga, E., *Historia pragmatica e interna de la universidad de Salamanca*, vol. 1 (Salamanca, 1914).

Fernández Alvarez, M., *Copérnico y su huella en la Salamanca del Barroco* (Salamanca, 1974).

Fernández – Carrión, M. and Valverde, J.L., *Farmacia y Sociedad en Sevilla en el siglo XVI* (Seville, 1985).

Fernández Duro, C., *Disquisiciones náuticas*, 6 vols., (Madrid, 1876–81).

Fernández Duro, C., *La Armada Invencible*, 2 vols., (Madrid, 1884–5).

Fournel-Guerin, J., 'La pharmacopée morisque et l'exercice de la médecine dans la communauté morisque aragonaise (1540–1620)', *Revue d'histoire maghrebienne*, **6** (1979), 53–62.

Galindo Antón, J., 'El arte de curar en la legislación foral aragonesa', *Actas III Jornadas Médicas Aragonesas* (Zaragoza, 1959), pp. 355–9.

García Ballester, L., *Historia social de la medicina en la España de los siglos XIII al XVI. La minoría musulmana y morisca* (Madrid, 1976).

García Carcel, R., *Herejía y sociedad en el siglo XVI. La Inquisición en Valencia 1530–1609* (Barcelona, 1980).

García-Villoslada, R., ed., *Historia de la Iglesia en España*, vol. 3 (Madrid, 1980).

Goodman, D., 'Philip II's Patronage of Science and Engineering', *British Journal for the History of Science*, **16** (1983), 49–66.

Gracia Rivas, M., 'El personel sanitario que participó en la jornada de Inglaterra. Nuevas aportaciones', *Revista de Historia Naval*, **1** (1983), 63–90.

Granjel, L. S., *La medicina española renacentista* (Salamanca, 1980).

Guilmartin, J., *Gunpowder and Galleys. Changing Technology and Mediterranean Warfare at Sea in the Sixteenth Century* (Cambridge, 1974).

Henningsen, G.,'El "Banco de Datos" del Santo Oficio. Las relaciones de causas de la inquisición española, 1550–1700', *Boletín de la Real Academia de la Historia*, **74** (1977), 547–70.

Herrera Puga, P., 'Enfermedad y prostitución en la Sevilla de los Austrias', *Actas, IV Congreso Español de Historia de la Medicina*, **1** (Granada, 1975), 125–133.

Herrero Hinojo, P. and Muñoz Calvo, S., 'Boticas y enfermerías en los monasterios Jerónimos', *Studia Hieronymiana*, **2** (1973), 465–84.

Hoffman, P. E., *The Spanish Crown and the Defence of the Caribbean (1535–1585)* (Baton Rouge and London, 1980).

Hossard, J., 'La pharmacie de l'Escorial: ce qu'elle fut, ce qui reste', *Revue d'histoire de la pharmacie*, **15** (1961–2), 134–9; 206–12.

Jara, A., *Tres ensayos sobre económica minera hispanoamericana* (Santiago de Chile, 1966).

Jiménez Muñoz, J. M., 'Salarios de médicos y cirujanos: Nóminas de Corte (1538–1600)', *Asclepio*, **33** (1981), 315–34.

Jordi González, R., *Relaciones de los boticarios catalanes con las instituciones centrales* (Barcelona, 1975).

Kamen, H., *Spain 1469–1714. A society of conflict* (NY, 1983).

Kamen, H., *Inquisition and Society in Spain in the sixteenth and seventeenth centuries* (London, 1985).

Kubler, G., *Building the Escorial* (Princeton, 1982).

Lea, H.C., *A History of the Inquisition of Spain*, 4 vols., (NY, 1906–8).

Loeb, I., 'La correspondance des Juifs d'Espagne avec ceux de Constantinople', *Revue des Etudes Juives*, **15** (1887), 262–76.

Lohmann Villena, G., *Las minas de Huancavélica en los siglos XVI y XVII* (Seville, 1949).

López Piñero, J. M., 'Valencia y la medicina del Renacimiento y del Barroco', *Actas del III Congreso Nacional de Historia de la Medicina*, **2** (Valencia, 1971), 95–108.

López Piñero, J. M., 'La disección y el saber anatomico en la España de la primera mitad del siglo XVI', *Cuadernos de Historia de la Medicina Española*, **13** (1974), 51–110.

López Piñero, J. M., *Ciencia y técnica en la sociedad española de los siglos XVI y XVII* (Barcelona, 1979).

López Piñero, J. M., 'The medical profession in 16th century Spain', *The Town and State Physician in Europe from the Middle Ages to the Enlightenment*, ed. Russell, A. W., (Wolfenbüttel, 1981).

López Piñero, J. M., Navarro Brotóns, V., and Portela Marco, E., *Materiales para la historia de las ciencias en España: s.XVI–XVII* (Valencia, 1976).

Lorenzo Sanz, E., *Comercio de España con América en la época de Felipe II*, 2 vols., (2nd ed. Valladolid, 1986).

Martz, L., *Poverty and Welfare in Habsburg Spain: The Example of Toledo* (Cambridge, 1983).

Matilla Tascón, A., *Historia de las minas de Almadén*, vol. 1: *desde la época romana hasta el año 1645* (Madrid, 1958).
Meerbeeck, L. van, 'Le service sanitaire de l'Armée espagnole des Pays-Bas à la fin du XVI^me^ et au XVII^me^ siècles', *Revue internationale d'histoire militaire*, **5** (1956), 479–93.
Menéndez Pelayo, M., *Historia de los heterodoxos españoles*, vol. 2 (Madrid, 1880).
Morata, N., 'Un catalogo de los fondos árabes primitivos de El Escorial', *Al-Andalus*, **2** (1934), 87–181.
Muñoz Calvo, S. *Inquisición y ciencia en la España moderna* (Madrid, 1977).
Muñoz Garrido, R., 'Empiricos sanitarios españoles de los siglos XVI y XVII', *Cuadernos de Historia de la Medicina Española*, **6** (1976), 101–33.
Muriel, J., *Hospitales de Nueva España*, vol. 1 (Mexico, 1956).
Muro, L., 'Bartolomé de Medina, introductor del beneficio de patio en Nueva España', *Historia Mexicana*, **13** (1963–4), 517–31.
Navarro Brotóns, V., 'Contribución á la historia del copernicanismo en España', *Cuadernos Hispano-Americanos*, **283** (1974), 3–24.
Olesa Muñido, F., *La organización naval de los estados mediterraneos y en especial de España durante los siglos XVI y XVII* (Madrid, 1968).
Oliver Rubio, F. and Zubiri Vidal, F., 'Un códice del siglo XVI', *Archivos del Estudios Médicos Aragoneses*', **4–5** (1957), 271–94.
Parker, G., *The Army of Flanders and the Spanish Road 1567–1659* (Cambridge, 1972).
Parker, G., 'Some Recent Work on the Inquisition in Spain and Italy', *Journal of Modern History*, **54** (1982), 519–32.
Parrilla Hermida, M., 'Un título de cirujano en 1579', *Asclepio*, **25** (1973), 173–8.
Parrilla Hermida, M., 'La anexión de Portugal en 1580: el hospital de campaña', *Asclepio*, **28** (1976), 275–8.
Pascarella, F., 'Un documento inédito del siglo XVI sobre el nombramiento del cirujano mayor del ejercito español en la Italia septentrional', *Archivo Iberoamericano de Historia de la Medicina*, **8** (1956), 315–20.
Pierson, P., *Philip II of Spain* (London, 1975).
Reglá Campistol, J., *Felip II i Catalunya* (Barcelona, 1956).
Ribera, J., and Asin, M., *Manuscritos árabes y aljamiados de la Biblioteca de la Junta* (Madrid, 1912).
Rivero, C. del, 'El ingenio de la moneda de Segovia', *Revista de Archivos, Bibliotecas y Museos*, **38** (1918), 20–31; 191–206; and **40** (1919), 146–50.
Rodríguez, A., 'El pronóstico astrológico que de Felipe II hizo el Doctor Matias Haco', *La Ciudad de Dios*, **96** (1914), 282–90; **97** (1914), 191–9; 364–72; 441–53.
Rodríguez Marín, F., *Felipe II y la alquimia* (Madrid, 1927).
Rudolph, W. E., 'The Lakes of Potosí', *The Geographical Review*, **26** (1936), 529–54.
Ruiz Martín, F., 'Las finanzas españoles durante el reinado de Felipe II', *Cuadernos de Historia*, **2** (1968), 109–73.
Sánchez Canton, F., *La librería de Juan de Herrera* (Madrid, 1941).
Sánchez Tellez, M., *et al.*, *La doctrina farmacéutica del Renacimiento en la obra de Francisco Hernández, c. 1515–1587* (Granada, 1979).
Santander Rodríguez, T., 'La creación de la catedra de cirugía en la universidad de Salamanca', *Cuadernos de Historia de la Medicina Española*, **4** (1965), 191–203.
Sicroff, A., *Les Controverses des Statuts de 'Pureté de Sang' en Espagne du XV^e^ au XVII^e^ Siècle* (Paris, 1960).
Simon Díaz, J., 'La Inquisición de Logroño (1570–1580)', *Berceo. Boletín del Instituto de Estudios Riojanos* (Logroño), **1** (1946), 89–119.
*Solemne sesión apologética celebrada por la facultad de medicina de Valencia para honrar la memoria de sus antiguos catedráticos los doctores Plaza, Collado y Piquer* (Valencia, 1895).
Somolinos D'Ardois, G., *Vida y obra de Francisco Hernández* (Mexico, 1960).

Taylor, R., 'Architecture and Magic. Considerations on the *Idea* of the Escorial', *Essays in the History of Architecture presented to Rudolf Wittkower*, ed. Fraser, D. *et al.*, (London, 1967), pp. 81–109.

Thompson, I. A., *War and Government in Habsburg Spain 1560–1620* (London, 1976).

Toribio Medina, J., *Biblioteca Hispanoamericana (1493–1810)*, vol. 2 (Chile, 1900).

Ulloa, M., 'Unas notas sobre el comercio y la navegación españoles en el siglo XVI', *Anuario de Historia Económica y Social*, **2** (1969), 191–237.

Ulloa, M., 'Castilian seigniorage and coinage in the reign of Philip II', *Journal of European Economic History*, **4** (1975), 459–79.

Ulloa, M., *La hacienda real de Castilla en el reinado de Felipe II* (Madrid, 1977).

Valverde, J. and Llopis González, A., *Estudio sobre los fueros y privilegios del antiguo Colegio de Apotecarios de Valencia* (Granada, 1974).

Vazquez de Prada, V., 'La industria siderurgica en España (1500–1650)', *Schwerpunkte der Eisen gewinnung und Eisen verarbeitung in Europa 1500–1650*, ed. Kellenbenz, H., (Cologne, 1974), pp. 35–78.

Vega y Portilla, J. de la, 'La botica real durante la dinastica austriaca', *Anales de la Real Academia de Farmacia*, **12** (1946), 349–406.

Vigón, J., *Historia de la artillería española*, vol. 1 (Madrid, 1947).

Vincent, B., 'La peste atlantica de 1596–1602', *Asclepio*, **28** (1976), 5–25.

# Index